MBOZI AREA

Forest Reserve

Alienation

NA   Former German Alienation, Returned to Native Authority

Revoked Alienation

—··— International Boundary

——— Area Boundary

— — Division Boundary

············· Village Development Committee Boundary

——— Main Road

— — — Local Road

◯ Area Headquarters

◯ Primary Courts

◯ Other Settlements

M.S. Gazetted Minor Settlement

† Missions

Sources: Base Map Compilation   Geological Survey of Tanzania, Quarter Degree Sheets 226, 227, 228, 242, 243, 244, 257, 258, (1:125,000)

Area Boundaries   Mbeya District, 1:500,000 (1962) and manuscript maps, Regional Surveyor, Mbeya

Local Political Boundaries   Manuscript maps, from field reconnaissance (1:50,000) June to October, 1967

Alienations   Manuscript maps, Regional Surveyor, Mbeya

Forest Reserves   Manuscript maps, Regional and District Forestry Officers, Mbeya

ECOLOGY AND CHANGE

Rural Modernization in an African Community

ECOLOGY AND CHANGE

Rural Modernization in an African Community

C. Gregory Knight

Department of Geography
The Pennsylvania State University
University Park, Pennsylvania

ACADEMIC PRESS
New York   San Francisco   London
A Subsidiary of Harcourt Brace Jovanovich, Publishers

ACADEMIC PRESS, INC.
111 Fifth Avenue, New York, New York 10003

United Kingdom Edition published by
ACADEMIC PRESS, INC. (LONDON) LTD.
24/28 Oval Road, London NW1

Library of Congress Cataloging in Publication Data

Knight, C   Gregory.
    Ecology and change.

    Includes bibliographical references.
    1.   Tanzania–Rural conditions.   2.   Agriculture–
Tanzania.   3.   Social change.   4.   Tanzania–Social
conditions.   I.   Title.
HN814.T32K57   309.1′678′04   73-7446
ISBN 0–12–785435–5

To Diane, Colin, and Corinna

Contents

Preface

The population of tropical Africa is predominantly rural. Even optimistic estimates of African development grant that the majority of her people will remain so in the foreseeable future. In fact, development economists suggest that the impetus and capital for development must come from the agricultural sector of the economy (Johnston and Mellor 1960; de Wilde 1967, 1:3).[1] During the colonial era in Africa, agricultural development was synonymous with introduction of cash crops. Although some attention was given to the problem of shifting cultivation and the destruction of natural resources that seemed to result from this practice, greatest official concern was with the introduction of cash crops and the devising of means by which these crops could reach international markets. Indirectly, the traditional food economy was also being altered. Population growth followed

[1]Simplified citations of references in the text refer to the Reference List, pp. 277–294.

pacification and introduction of medical services, meager though they remain. Cash crops required labor that might otherwise have been available for food-crop production and, in areas of population pressure, took land from the traditional food-crop cycle of production and lengthy fallow. Hence, tropical Africa today is faced with the dual problem of continued specialized cash-crop production as the foundation for growth of the whole economy and adequate feeding of its growing population. It is indeed a paradox that the predominantly agricultural societies of Africa should become short of food (Steel 1965; Gardiner 1968).

This is a geographical study of agricultural practices and agricultural change focused on one society of East Africa, the Nhiha[2] of Mbozi Area[3] in southwestern Tanzania (see front endpapers). Mbozi is particularly attractive for study because it provides an intriguing blend of cultural diversity, differing population densities with concomitant stages of agricultural intensification, varying levels of modernization, and a spatial structure of resources representing a variety of agricultural potentials, as well as a number of degrees of accessibility or remoteness. Potential as well as actual developments in the international accessibility of Mbozi suggest that the processes of change will hasten in the near future.

There have been many case studies of traditional agriculture in Africa (Trapnell and Clothier 1937; Trapnell 1943; Richards 1939; Peters 1950; deSchlippe 1956; McMaster 1962a; Porter 1965) as well as studies of agricultural development (Clayton 1965; Kay 1965; de Wilde 1967).[4] This work is of neither genre; rather, it is a contribution intended to explain the traditional agricultural system of a particular society and the process and results of agricultural change within it. It is of particular importance because most of the change that has taken place in Mbozi has not been the directed change that has been the focus of much agricultural

[2]The practice of using prefixes on the noun root referring to a people is followed here. For example, the land area occupied by the Nyiha is "Unyiha" and their language, "Shinyiha." The usual prefix used in reference to the people themselves is dropped for simplicity. Thus, I use the term "Nyiha" as synonymous with "Wanyiha."

[3]Under the British, Tanganyika was organized into Provinces and Districts. Since independence in 1961 and subsequent union with Zanzibar to form the United Republic of Tanzania in 1964, the term "Region" has replaced "Province," and the former "Districts" are now "Areas." Mbozi Area was formerly part of Mbeya District but was separated in 1964; it is incorporated within the larger Mbeya Region.

[4]Bibliographies of agricultural development projects in Africa have been compiled by Freitag (1963) and Neville-Rolfe (1969).

development literature. In Mbozi, change has resulted from a voluntary learning experience that may suggest strategies for development where funds for personnel and massive programs are not available.

As used in this study, the phrase "traditional agriculture" will refer to agricultural practices used at the time of European contact. It involves both the way in which a people classify or evaluate the environment they occupy and the way in which they utilize that environment for production of needed commodities. Traditional agriculture was by no means static. As we shall see in Chapter 6, changes were common in precolonial as well as modern time. The longer a people occupy an environment, the more they learn to interpret and use it with greater facility. They continue to cope with and seek solutions for the many uncertainties they face—weather, pests, and diseases. In time their understanding approaches, within their cultural framework, conclusions that our culture would term "scientifically valid" (Blaut 1961; Knight 1971a). In addition to the greater understanding that came from interaction of man and environment, contact with other cultures through migration and spread of new crops and ideas brought change in the traditional agricultural sphere. For example, maize and cassava spread rapidly after their introduction into Africa by the Portuguese (Jones 1957, 1959; Miracle 1958, 1965, 1966). Although traditional agriculture was dynamic, it was usually organized for the production of food and native industrial crops on a subsistence level —each family or production unit raising little more than what it, itself, would consume. The traditional sphere, with its environmental interpretations and agricultural techniques, persists today in many, if not most, societies of tropical Africa. In some, traditional agriculture has been altered almost beyond recognition (the Kikuyu of Kenya: Clayton 1965), while among other peoples, including the Nyiha, it remains more or less intact. In the latter case, cash cropping is a veneer under which traditional practices are more slowly evolving (de Wilde 1967, 1:4).

Traditional agriculture throughout tropical Africa has primarily been of the shifting or slash-and-burn variety. This term encompasses a range of agricultural systems and techniques.[5] In all of these, however, land is cleared, cropped for a limited time, and abandoned to an uncontrolled fallow of lengthy duration as a means for restoring the system's productivity. Thus, the farmer's attention *shifts* from field to field

[5]Morgan (1969) has mapped and discussed this variety of traditional agricultural techniques in tropical Africa.

through the years. More rarely, he shifts his homestead as well in search of land ready to be cleared for cultivation. A considerable literature has accumulated on shifting cultivation (Conklin 1961). Scientific assessment of this agricultural type has changed from total condemnation to recognition of an inherent ecological rationality in the system. Shifting cultivation makes agricultural production possible in areas of generally infertile soils subject to rapid depletion of plant nutrients due to high temperatures and moisture (Bartlett 1956; Gourou 1956). There are notable exceptions to shifting cultivation in the more densely populated areas of Africa, the significance of which will be mentioned presently.

Historical study indicates that shifting cultivation has been an early method of agricultural production in virtually all forested areas of the world when subsistence production supported a relatively sparse population. Shifting cultivation was common in Europe and in Scandinavia, in particular, into the twentieth century (Clarke 1947; Davies 1952; Mead 1953; Montelius 1953). It was equally common in colonial America where early settlers, in response to plentiful land resources, reverted to traditional American Indian techniques of slash and burn (Carrier 1923; Gras 1925; Sauer 1941; Waibel 1950).

The most intriguing model of a kind of universal scale of agricultural change has been formulated by Ester Boserup (1965). Her scheme will be discussed in Chapter 7, but it should be noted here that in the Boserup formulation, as population increases, the intensity of land use increases with shortened periods of fallow and increasing frequency of cropping. Most societies, according to Boserup, practice techniques potentially suitable to continuous production on a plot of land. The continuously cropped garden plot is widespread in most African societies, for example. Thus, as population grows there is an evolution in emphasis from extensive to intensive techniques. These kinds of intensive techniques support dense populations. The existence of dense populations in areas lacking superlative environmental resources lends support to Boserup's theory.

In contrast to Boserup's rather optimistic view of population growth and agricultural change, the work of William Allan (1965) is derived from his experience with a particular variety of shifting cultivation found in Zambia, the *citemene* system.[6] Allan suggests that overpopulation is a real threat to the continued viability of an agricultural system. His thesis, discussed more fully in Chapter 7, states in essence that given a certain length of

---

[6]The former spelling of this term is *chitemene*. Here, more recent convention is followed with the spelling *citemene*.

fallow required to regenerate sufficient vegetation to maintain a slash-and-burn system, and given a certain amount of land required to be cultivated per person per year, there exists an equilibrium that represents the maximum population density the system can support. From this, we can suggest that if the population density exceeds the critical value, the fallow must be shortened. But, having shortened the fallow, land is then less productive and the area in cultivation in any one year must increase, which causes a further shortening of the fallow. Hence, a spiral results that seemingly ends in disaster. The underlying premise here is that techniques for improving productivity or longevity of land use are not available.

Perhaps intensive techniques may have been lacking in some areas; but more commonly, intensive techniques are known. In Mbozi, the latter has been the case, and we will argue that Boserup provides a reasonable working set of assumptions from which to build a model of agricultural change. We will be interested in evaluating agricultural changes in the face of increasing population, especially in view of the fact that some of the most productive land of Mbozi has been planted to coffee, a permanent crop, in effect increasing the population density further so long as importation of foodstuffs does not take place. Then, we shall make a transition from local population pressure on resources to the spatial and temporal patterns of rural development in Mbozi.

Mbozi, like other areas of rural Africa, has increasingly experienced a process of rural modernization.[7] In rural Africa, we see obvious manifestations of development toward a modern world model of human activity, ranging from political behavior to factory-made hoe blades to textbooks. But underlying these surface artifacts of the modern world is a fundamental evolution of personal and aggregate commitment to a radically different scale of life from that once extant. This scale is primarily the typing of the individual, family, and local society into the regional and national mosaic of institutions, activity, and belief. This commitment may evolve within rural societies as they actively, selectively, and voluntarily alter their own behavior and beliefs; it may be forced upon them by their involuntary relationship with the colonial power or by their contemporary incorporation in a modern state. There are many fibers that constitute the fabric of modernization into which rural areas are tied.

---

[7]The notion of modernization as a multidimensional transformation of traditional societies was elaborated by Lerner (1958). Inkles (1970) and others have discussed the modernization concept from the viewpoint of individual attitudes.

Rural societies become incorporated into the institutions and activity of the regions, nations, and even the world within which they are embedded. Economically, they become dependent upon the market for consumable commodities, for production inputs, for products once manufactured locally, and for disposal of produce. This dependence often goes beyond the market itself to include institutions of the centralized government tied to rural locales by the upward flow of funds in the form of taxes and a downward flow in the form of agricultural extension, experiment stations, and marketing systems. Politically, rural areas become incorporated in larger institutions. In Tanzania, for example, the contemporary political structure links individuals and households to the national government. In addition to the centralized decision making, taxation, and allocation of development resources implied by this linkage, political party structure provides a similar hierarchical structure in the ideological realm. Communication and transportation systems provide channels for flow of people and beliefs as well as commodities. These networks evolve both as a stimulus for and response to modernization, with rural areas becoming increasingly accessible or linked to other areas and to local, regional, and national urban centers.

Migration among rural areas and between country and city results from and also creates a flow of information which ties regions together. While the motivation to migrate is complex—ranging from temporary target labor to provide a financial stake for investment or bride-price, to a permanent commitment to a new area—this flow has its reactions in both source and destination areas. At the source, for example, altered agricultural systems evolve to cope with the absence of large numbers of men; rural populations become increasingly sophisticated as migrants return with wider experience and increasing material demands. At the destination migrants create the well-known problems of uncontrolled urban growth typical of many African cities; in rural areas they may also interact in locally recognized cultural antagonisms.

Belief systems are altered in the process of modernization. Contact is made with the larger religious traditions; information flows through national radio stations, newspapers, and magazines; the iconography of national culture is disseminated; Western-based educational systems alter traditional aspirations and world views; and motivations are created through activity within modernizing institutions.

These selected dimensions of modernization evoke the nature of this fabric of change. Thus, a major purpose of this monograph is to outline the process of rural modernization in Mbozi. Our perspectives are broad, perhaps broader than fully justified by the data I gathered there. Nevertheless, the viewpoints we shall discuss outline an approach toward understanding change from a holistic viewpoint, suggesting many directions for subsequent, more specifically focused research among the Nyiha. These viewpoints include the basic ecology of agronomic systems set into a regional ecological systems; the ethnogeography or folk geography of the people involved; the spatial organization of society and landscape developing through time; and the political, social, and economic matrices in which the evolution of the Mbozi economy, society, and landscape is embedded. Each of these perspectives requires some introductory comment.[8]

The ecological point of view causes us to focus on scientifically understandable processes in the environment, including the very basic mustering of energy resources for life. Data presented in this study are only indicative of the basic flow of matter and energy that composes any ecological system. Human activity channels available environmental resources (solar energy, water, soil nutrients, stored nutrients in vegetation) through harvestable ecological entities (crops, livestock) into the basic sustenance for human society. In a sense, we treat ecological process and the landscape resulting from it as an environmental absolute. People must cope with and adapt to that environment to produce their needs; the process of adaptation may itself alter the environment in ways requiring further adaptation.

The ethnogeographic point of view asks that we treat the cognitive component of this human mustering of environmental resources as a culture-specific system of knowledge and beliefs concerning the environment and its human organization and use. Predicated on the assumption that behavior is mediated by cognitive processes, ethnogeography (as ethnoscience in anthropology) seeks to understand human behavior, in this case with respect to resources and the spatial environment, in relation to shared systems of cognition that constitute portions of a culture. In a preliminary sketching of some elements of the ethnogeography of the Nyiha, we find that these elements can be cross-culturally mapped or compared with our system of knowledge (science) and evaluated. Although my preliminary queries

---

[8]These viewpoints are related to those proposed by Firey (1960): the ethnographic, ecologic, and economic perspectives on human resource use.

in this area could not hope to reveal the total Nyiha system of cognition, we can find examples where this traditional knowledge may have greater empirical validity than knowledge from our own system. In Mbozi, to plant a crop early is to create a risk that need not exist. If one plants finger millet, for example, in October or even November, he risks the break in the early rains that inevitably comes in November or December before the continuous rains occur. If the soil has not been fully wetted, in even a short break of a week or two, soil moisture within reach of seedlings may be depleted and young shoots wilt beyond recovery. If, on the other hand, a farmer waits until December or January, he is virtually guaranteed a good harvest. When seed is scarce, with delayed planting the farmer can be sure that the rains have started. Were yields to be reduced by late planting, this would be a calamity much less significant than possible loss of the seed (Wrigley 1961:84). Looking at the other end of the season, were a farmer to plant early, then the finger millet would come to maturity before the end of the rains. Nyiha experience is that a rust or mildew is almost inevitably the result. In Mbozi, agricultural extension officers extolled the virtues of early planting as a means to exploit early nitrogen flushes in the soil as well as to enable the crop to realize the benefit of the total seasonal rainfall. They had failed to observe, however, the local rationale for ignoring this advice. Focused inquiries could have elicited the reasons for people's valid refusal to plant early!

The spatial organization of society and landscape includes such elements as population density, settlement patterns, land-tenure and land-use patterns, landscape units, transportation networks, patterns of human daily activity, and a myriad of other phenomena. It is these elements the geographer dissects from the apparent confusion of the region in order to discover the elegant underlying organizational principles which, when recombined, lend organization and rationality to the landscape. Time is a crucial element here. Whether a glimpse of the landscape comes through aerial photography, motor vehicle tour, or foot reconnaissance, any such glimpse provides only a minimal slice through time. Land-use patterns, for example, reflect seasonal, rotational, and long-term evolutionary processes. Rather than simple correlations among activity, spatial organization, and environment, we seek techniques that allow us to document or adduce processes acting through time, so that our investigation, here and now, becomes enriched with a larger perspective. Thus, land-use patterns become artifacts of systems of agricultural activity operating in and evolving through time.

Finally, Mbozi is embedded in a spatial matrix of people and social activity that can be seen from political, social, and economic viewpoints. The matrix includes a local plural society consisting of indigenous and immigrant Africans, European planters, and Asian merchants; involvement in the process of rural modernization; development guided, yet limited by national Tanzanian aspirations and policy; and a world political and economic system seemingly beyond potential Tanzanian if not tropical African influence.

The chapters that follow result from initial geographical field research in Mbozi Area undertaken during 11 months in 1966 through 1967. Numerous interviews with individuals and small groups were conducted through happenstance, purposeful selection, and formal sampling for specific data generation. The medium of communication was the Tanzanian national language and local *lingua franca*, Kiswahili, or Shinyiha, the language of the major ethnic group in Mbozi, the Nyiha. The latter language was used through the aid of my capable interpreter, E. N. Sampamba. Terminology of Mbozi peoples, including the Nyiha, is rendered here as roughly phonetically equivalent to standard Kiswahili and is written in Kiswahili orthography for simplicity's sake. Linguists and others interested in Bantu languages and Shinyiha in particular will find suitable discussion in the works of Busse (1960), Greenburg (1963), and Guthrie (1967, 1969).

This study could not have been completed without the support of many individuals and institutions. At the risk of slighting some, special mention should be made in several cases. Field work in Tanzania (1966–1967) was supported by the United States National Science Foundation (Grant GS-1109), and subsequent data analysis was partially supported by a grant from the University of Kansas General Research Fund. The University College at Dar es Salaam made field study possible through my appointment as Research Associate. Staff members L. Berry and I. D. Thomas made significant logistical contributions. The Tanzanian Ministry of Agriculture, Forests, and Wildlife provided official endorsement, and through their Mbeya Region and Mbozi Area offices provided invaluable support. The Mbozi Agricultural Experiment Station cooperated fully with collection of basic environmental data. Maps of Mbozi Area and this report can only partially repay the opportunity provided for research by the Tanzanian government and university.

Local field research was facilitated by the Minister of Regional Administration and officials in the political hierarchy from the Minister himself through innumerable *Kumi-kumi* chairmen. Constant encouragement and cooperation of missionaries, the Mbozi Farmers' Association (an organization of estate owners and managers), and the Asian trading community made possible investigation of the processes of change from several points of view. The excellent efforts of my interpreter, E. N. Sampamba, and field assistants, L. N. Silwimba and S. K. Kajula, is gratefully acknowledged. Thomas Nkota contributed in strong measure to the successful logistics of the field session. My wife, Diane S. Knight, completed time-consuming gravimetric determinations of soil moisture content throughout the research year.

I am also indebted to the East African Meteorological Department for reproducing climatological data; to the East African Herbarium for identification of plant specimens; to the University of Minnesota, The University of Kansas, and the Pennsylvania State University for grants of computer time, and to Geoffrey Roper for his aid in computation; to my students at Kansas and Penn State with whom many ideas were tested; to P. W. Porter for his many contributions; and to Diane S. Knight and Beverley Brock who provided invaluable criticism of this document in various stages of its preparation. Finally, there are few words that could express my debt to and admiration of my Nyiha friends.

Mbozi

chapter 1

"Watamanyile ukuwuzilizya numo atakumuwuzya
nazimo."–
He who does not ask will not be asked anything.
NYIHA PROVERB (Busse 1960:132)

Arrival of the bird *umwanga* announces the beginning of *mundundu*, season of the early rains.[1] The plateau surface again takes on the lush tropical green that will last through *ishisiku*, the rainy season. Where virgin or secondary woodland remains, Nyiha men have finished cutting trees to make way for new fields. The month of nights broken by the glow of flaming woodland has ended, and the family's effort turns to the hoeing of burned fields in preparation for planting. Fallow grassland is turning from dry grass beige to the green of emergent growth, and finally to the brown of freshly turned soil. Women undertake much of this hoeing, for coffee calls for the labor of men. Work in coffee planting is rewarded now by the delightful aroma of coffee trees in flower (not unlike the scent of jasmine, honeysuckle, or citrus) as well as the expectation of eventual harvest

[1]See Table 1.1 and front end papers.

Table 1.1

Nyiha Seasons

Season	Western months	Related terminology
	January	
Ishisiku		*Ishisiku:* the rainy season
	April	
	May	
Ulupepo		*Ulupepo:* cold season *Pepo:* wind, season of shift from southeasterly to northeasterly winds
	July	
Ishisanya		*Ishisanya:* dry season *Ulusanya:* sunshine or heat
	October	
	November	
Mundundu		*Mundundu:* early rains *Umulundu:* the first rain
	December	

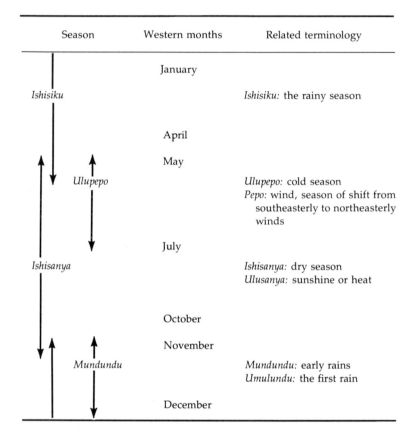

of this year's coffee crop. Hoeing will extend for several months during which the major grain crops—finger millet and sorghum—are planted. Women will continue to tend and weed family food gardens throughout *ishisiku;* men will labor on their own fields, in coffee plantings, or perhaps in wage labor on a coffee estate.

As the rains begin to fade, major crops have matured toward harvest. *Ishisanya,* the dry season, is introduced by the flowering of *impangwe* indicating arrival of the cold period of *ulupepo.* Citrus and bananas in valleys may be tinged by frost, for even a tropical cold season at an altitude of 5000 feet is marked. Grains are harvested and dried for storage; coffee is carefully hand-picked and processed for sale through the cooperative. For the majority without steady wage labor, products of the *duka* become accessible only through sale of coffee or a portion of the food harvest. Such staples as salt, sugar, or cooking oil; clothing; farm implements; or further saving for plow and oxen, radio, or bicycle— all compete for scarce financial resources.

As the dry season progresses, the landscape becomes parched; trees lose leaves; and grasses dry as do seasonal streams. Young boys herding cattle must take them further to permanent water sources. Lush grass of the *mbuga* seasonal swamp turns dry and unpalatable; it will be cut, however, for thatching houses. Much grassland will be burned to encourage fresh growth for cattle. Trees and saplings will be cut for new fields or for building frames held together by strips of bark, later to be covered with clay. At each homestead, craft activities undertaken during spare moments during the cropping season will now occupy more time: women making and baking clay pots; men weaving mats and baskets, carving new implement handles, shaping musical instruments. Music and dance receive more time, with many a night remaining alive with the sound of drums and conviviality. Beer is always plentiful after the millet and sorghum harvest.

The sun again comes high into the sky toward the end of *ishisanya* bringing warmer weather whose heat potential will eventually be tempered by the cloudiness accompanying the rains of *ishisiku*. The daily tempo of the cropping season begins to reemerge—early rising for work in the fields, returning in the middle afternoon for food, work, and relaxation around the homestead. As always, the homestead is alive with people and livestock. Chickens scurry from underfoot; dogs announce the arrival of the stranger who will certainly receive hospitality; boys herd sheep, goats, and cattle during the day, bringing them home to be penned for the night. Young children of many mothers and one father play together; older children arrive home after a day at school. Water is carried to the homestead in pots or gourds, and staples are prepared for the major meal of the day.

Umwanga arrives, and cooling rains again paint the landscape green. A new season begins; season follows season; children grow, marry, and die; families add the symbols of the modern world to their possessions; Christian churches beckon and receive converts; paths between separated homesteads deepen with use; other paths become vehicle tracks, while still others evolve as local highways annually hoed clear of weeds by a crew paying labor for taxes. Woodland cut is fallowed for too short a period to regrow, and the landscape becomes a man-generated grassland punctuated only by fruit trees.

Men working on coffee estates grow their own coffee, and as one neighbor learns from another, coffee growing diffuses across the landscape, beyond its seeming ecological limits. The cylindrical Bantu house ubiquitous in 1950 gives way to an almost universal rectangular style within two decades. The commitment is made, perhaps unknowingly, to dependence upon a system of life larger in scale than the local society, and well beyond its control. Petty chiefs become village headmen under the British; some, but not all, emerge as local political leaders after independence. TANU flags adorn many of the one-in-ten houses of the *Kumi-kumi* cell leader.

Daily bus service connects Mbozi with the regional city, Mbeya, 50 miles distant. Most men will travel at least once outside the local district in search of employment and learn Kiswahili. Women, being less migratory, have little need for Kiswahili, and most remain conversant only in the local tongue. A growing cadre of young adults are literate, attesting to the role of an expanding network of schools. National and international culture, music, news, and information become increasingly available as wage labor or cash cropping provide a means for purchase of radios. European estate owners and managers provide opportunities and models for change whose importance far exceeds the proportion of the landscape they control. Asian merchants and aspiring local entrepreneurs offer the wares of the world. The Great North Road to Zambia and the Copper Belt, to Kenya, and to Dar es Salaam will shortly be paved through the district. The railroad connecting Dar es Salaam with the Copper Belt will soon follow the general route of the recently built petroleum pipeline.

The rains and seasons retain their cyclical nature; life maintains its rhythm of birth, maturity, and death; but change progresses, compounds, and accelerates. This is Mbozi, Tanzania, in the late 1960s. Our task is to freeze her for a moment in time; dissect and analyze her; and extend our analysis backward in time and outward in space, eventually to reassemble and understand her, perhaps to catch a glimpse of her future, perhaps simply to learn of her evolution.

The Corridor Region

In the region between Lake Tanganyika and Lake Nyasa,[2] highland areas adjoining the Western and Eastern Rift Valley systems of East Africa merge to continue south as the Malawi and Rhodesian highlands. Broken only by isolated mountain peaks, stream dissection, and the trough of the Zambezi River, this series of highlands and plateaus is the only moderate-to high elevation route from North and East Africa to the southern portion of the continent (Figure 1.1). Marked altitudinal differences in temperature, rainfall, and vegetation within short distances suggest a multiplicity of agricultural resources and related land-use potentials.

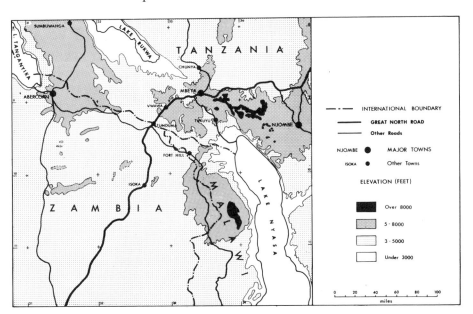

Figure 1.1. Tanganyika–Nyasa Corridor.

[2]The traditional names for these lakes have been used here in the absence of any international convention to revise them. Some maps do indicate Lakes Tanzania and Malawi rather than Tanganyika and Nyasa.

Monica Wilson (1958) coined the term "Nyasa–Tanganyika Corridor" for this general region in which Mbozi is located. This name reflects the role of the particular configuration of landforms and lakes in funneling people and ideas between eastern and southern Africa. The Corridor may have been important as the major route free of the tsetse fly and animal sleeping sickness through which cattle, the idea of cattle husbandry, and milking passed to southern Africa (Murdock 1959). Bantu migrations in the area are still not fully understood, but successive movements from a core area in the southern Congo into this region have been established (Huntingford 1963; Wills 1964). Ngoni peoples, related to the present Nguni culture of Natal, South Africa, migrated northward in the early 19th century. They reached the Zambezi by 1835 and occupied Ufipa in 1845, eventually extending as far north as Lake Victoria. Arab slaving and ivory trade routes crossed the Corridor from 1850 until the suppression of slaving in the decades following the British treaty with the Sultan of Zanzibar in 1873. Raids of the Ngoni, Arabs, and others involved in the ivory and slave trade (the Bemba, for example) resulted in disruption and dispersal of Corridor residents. Related peoples became separated as numerous groups left traditional areas to seek refuge in the dissected mountain areas of the Corridor. There, having assumed new names based on location or names of leaders, they remain as separate cultural entities, isolated peoples whose former cultural affinities have not been fully established.

With its central location in the Corridor, Mbozi lies at the point of intersection of several different agricultural systems that occur within portions of the region. The *citemene* system of shifting cultivation of finger millet and sorghum occurs throughout Zambia. In this technique, trees cut from a wide area are gathered in one large or several small circles and burned, resulting in distinct circular plots. Recently, cassava has been added to some *citemene* systems. In Malawi, maize is grown on small mounds. As the garden is cultivated, weeds and soil are heaped into mounds between those presently cropped and left to compost. These weed–compost mounds are used for the following crop, and new weed mounds are hoed for the subsequent year. In southwestern Tanzania variants of a different mounding system also occur, in which grassland mounds are used as a soil-preparation technique for finger millet. And, finally, in northern Zambia and southwestern Tanzania, there is a more general type of shifting cultivation in which the whole cut area is burned and cropped. Thus, one fascinating potential of research here was to sort out, environmentally and culturally, these agricultural traditions, a task at least partially completed. Indeed, Mbozi fulfilled its promise of being intriguing both environmentally and culturally.

Mbozi remains in a critical Corridor location in spite of its isolation from the major Tanzanian and international markets focusing on Dar es Salaam. In 1967 about one-half of the 600 miles to Dar es Salaam was traversed by the *murram* (dirt) Great North Road. Hard surface road or railroad provided transportation for the remaining 300 miles. The major market on the Zambian Copper Belt was also 600 miles distant via *murram* road. Political events of recent years resulted in a focus on the route from the Copper Belt to Dar es Salaam, passing through Mbozi. The oil lift of the late 1960s to Zambia from Tanzania (a result of disruption of transportation routes through the Congo, Angola, and Mozambique) brought attention to the totally inadequate state of the Tanzania–Zambia linkages for heavy haulage. In 1967 plans were under way to upgrade the Great North Road to a bituminous surface; to build a petroleum pipeline through Mbozi linking Zambia with Dar es Salaam, with local terminals en route; and to build the Tan–Zam Railroad from the Copper Belt through Mbozi to Dar es Salaam. By 1973 the first two projects were completed, and construction of the railroad had reached the Zambian border.[3]

Mbozi Area

The administrative center of Mbozi Area is at Vwawa in the heart of Unyiha, home of the Nyiha, on the Mbozi Plateau proper. The plateau extends as a peninsular northwestward-dipping block into the rift valley of Lake Rukwa, and is bounded on the southwest by the Msangano Trough and its extension along the Nyasa Songwe River to Lake Nyasa, and on the northeast by the Songwe Trough, with the Rukwa Songwe River drain-

[3]See O'Connor 1965 and Griffiths 1968.

ing into Lake Rukwa. Lake Rukwa is a remnant of a larger lake that once covered the two troughs, leaving behind recent lacustrine sediments that characterize lower portions of the troughs and the Rukwa basin (Figure 1.2). Most of Mbozi Area drains to Lake Rukwa and is part of its basin of interior drainage. Above the Msangano rift wall (known locally as the Chingambo Hills) in the southwestern part of Mbozi Area lies a portion of the Ufipa Plateau, home of the Namwanga who extend into Zambia as well as into the Msangano Trough. Other peoples

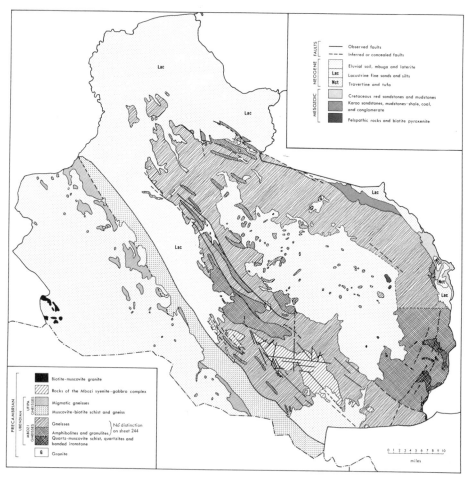

Figure 1.2. Geology of Mbozi Area. [Compiled from Geological Survey of Tanzania, Quarter Degree Sheets 226, 227, 228, 242, 243, 244, 257, 258 (1:125,000).]

in Mbozi Area are the Wanda who live in the Rukwa Plain, the Lambya and Malila who occupy the Mbozi Fault Block Hills, and the Songwe who live in the Rukwa Songwe Valley.

Both the Mbozi and Ufipa Plateaus are composed of Precambrian crystalline rocks. The central part of the plateaus has a rolling topography with a scattering of conspicuous mountains, isolated erosional remnants standing up to 500 feet above the surface (Figures 1.3 and 1.4). The edges of the plateaus are stream dissected and much of the land is in steep, rocky slopes. The major fault zones bounding the plateaus have a refief of 400–800 feet (Figure 1.5). However, the Mbozi Plateau dips toward the northwest disappearing beneath the lacustrine deposits of the Rukwa basin. On the poorly-drained upper plateau surfaces, the landscape is broken by grassy seasonal swamps known in Kiswahili as *mbuga,* in the Nyiha language as *ihombe,* and by such terms as *dambo* or *vlei* in southern Africa. The well-drained uplands were once covered with *Brachystegia* or *Miombo* woodland, much of which has now been cut for agricultural purposes. The soils of the area vary in quality. The best soils are those of the Mbozi Plateau and the hills to the east. These are derived from air-borne volcanic material from the Rungwe Volcanoes ringing the northern end of Lake Nyasa (Harkin 1960). Elsewhere in the uplands, lateritic soils vary in fertility depending upon their degree of active erosion. Rather droughty alluvial soils characterize the troughs and the Rukwa basin (Spurr 1955; P. Brock 1963).

Elevation of the Mbozi Plateau is from 4500 to 5500 feet; therefore, temperatures are subtropical rather than tropical. Daily highs may reach the mid-80s while morning lows are often a crisp 50 degrees. On the plateaus, frosts are not altogether rare, and bananas, for example, are seldom planted in the bottoms of steep valleys subject to cold air drainage. Fülleborn (1906:477) stated that temperatures of $-4°C$ (25°F) were recorded in Unyiha. Rainfall is highly seasonal, with the Mbozi Plateau average of 50 inches concentrated in the months from October to May. Rain in June, July, or August is a rare event. General air flow from the southeast brings rain from the Indian Ocean with Lake Nyasa as a secondary moisture source. Amounts of rainfall received decrease rapidly toward the northwest. Mkulwe Mission in the Rukwa basin receives an average of only 32 inches.

Figure 1.3. Looking west from Ng'amba Mountain. The smaller hill Nasumoli stands in front of the Msangano Trough and, on the horizon, the Chingambo Hills. Note the cutting of *miombo (Brachystegia)* woodland for fields.

Figure 1.4. The hills bounding the Mbozi Plateau on the east. Taken along the Great North Road between Mlowo and Ruanda, this photograph shows extensive cutting of *miombo* woodland. What appears to be continuous woodland at eye level is in fact a landscape checkered with *nkomanjila* fields in cultivation or fallow regrowth.

Figure 1.5. The edge of the Mbozi Plateau from the Songwe Trough. The dissected plateau margin stands 400–800 feet above the trough, an arm of the Rift Valley system of East Africa.

While 50 inches is more than adequate for crop production at 5000 feet, 32 inches is marginal in the Rukwa basin at a warmer 3100 feet. Normal variations in precipitation from year-to-year in the Rukwa basin can bring crop failure and famine. Near Tunduma and Ndalambo, precipitation averages 40 inches.

Mbozi was a locus for both European and Asian colonization. The first settlers were German missionaries. After 70 years the impact of the first missionary, Traugott Bachmann, remains clear—his local nickname, *Mwalwizi*, is well known and linked to the crops he introduced to the Nyiha. European settlement and the arrival of the Asian merchant community followed discovery of gold at Lupa in Chunya Area (to the northeast of Mbozi) and the opening of land for settlement, both in the 1920s. Thus, elements of contact with non-African peoples and resulting change have been present for seven decades (Figure 1.6). Introduction of cash cropping and the Great North Road (Figures 1.7 and 1.8) initiated continuous contact with the modern world. The numerically dominant Nyiha, among whom Europeans and Asians settled on the productive Mbozi plateau, continue to

Figure 1.6. The view from Ng'amba Mountain. Ng'amba Farm, a coffee estate, is visible in the center of the photograph. The dark areas are coffee plantings—in the foreground notice African farmsteads and fields. The Chingambo Hills fronting the Ufipa Plateau appear above the Msangano Trough in the background. Looking south, Malawi is at the left of the photograph, and Zambia at the center and right.

Figure 1.7. The former route of the Great North Road near Mbozi Mission. The German-built bridge across the Nkana could no longer be used by motor vehicles after 1958. However, during the paving and realignment of the Great North Road, this access route to the mission was restored.

Figure 1.8. The Great North Road near Ruanda, 1967. The wrecked truck was carrying petroleum to Zambia. This section of the road plus that climbing the Chingambo Hills near Tunduma were annually closed during the height of the rains. Hard surfacing begun in 1968 has improved the road considerably.

use traditional food-production systems combined with a veneer of wage labor or cash cropping. The Nyiha have expanded from their core area outward to occupy the wooded, dissected plateau margins, including the plateau's arid northwestern extreme; but this migration has failed to check emergent pressure of population on land resources in central plateau areas most ideal for coffee.

Within Unyiha, some 20% of the population are Nyakyusa or Ndali in origin, immigrants once attracted by employment opportunities in European coffee estates and by the availability of land in Unyiha compared to their very densely populated homelands. While the Songwe, Lambya, and Malila are also involved in the cash cropping economy, neither the Wanda nor the Namwanga have been incorporated into the modernization process to an equal extent. Only recently was Turkish tobacco introduced to the Ndalambo Namwanga as their first cash crop.

While the Msangano Namwanga and Wanda do produce sesame seed and cattle in exportable quantities, these areas lag considerably behind Unyiha in development, due both to poorer resources and inaccessibility.

Like people all over the world, the Nyiha of central Mbozi earnestly claim that their environment is the best for agriculture. Such a claim is understandable. Ample rainfall, truly *temperate* temperatures, seasonality of climate and vegetation, and adequate soils suggest that like other tropical highland areas, Mbozi is indeed a delightful place to live and farm. It was certainly an enticing place for research.

The People

chapter 2

"Inana ja kungulu jitakulesela ivala. Zye ujise akuvomba zye wope numwana akuvomba zizyo."– The crow cannot toss away the white mask given it by nature.
NYIHA PROVERB (Busse 1960:131)

The Nyiha are a Bantu-speaking people who arrived in Mbozi Area as part of large-scale migrations during the last millenium. Their traditions preserve legends of the arrival of ancestors of traditional chiefs who found in the area an indigenous hunting and gathering population possessing neither fire nor agriculture.[1] Although it is attractive to think of the founders of chiefly lines introducing fire and agriculture to a primitive people (Figure 2.1), these myths serve more to legitimatize traditional political leadership than to provide reliable history of the Nyiha and their cultural forebears (Brock 1968). The Tanganyika–Nyasa Corridor region is likely to have been occupied by Bushmen hunters and gatherers sometime prior to the arrival of the Bantu (Hunt-

[1]Nyiha chief Mwamlima referred to these indigenous peoples as Ukambombo or Wantandala (Mwamlima 1967).

15

Figure 2.1 The Nyiha *intuma*. Some legends assert that an aboriginal hunting and gathering people lived in this conical, sapling-framed, grass-covered house. Contemporary Nyiha use the easily built house for use by guests or as a temporary dwelling.

ingford 1963:92). However, archaeological remains at Kalambo Falls and Ivuna both provide evidence of centuries-long occupation of portions of the Corridor by agricultural peoples. Indeed, Phillipson (1968:211) has suggested that early iron age peoples brought pottery, metallurgy, and food production to Zambia in the early centuries of the Christian era.

The Kalambo site, near Lake Tanganyika in Ufipa, consisted of iron age villages occupied for some 800–1000 years after about 350 A.D. (Fagan and Yellen 1969). Remains at Kalambo included grindstones, bovid teeth, and beans, suggesting the early presence of agriculture (Seddon 1968). The Ivuna salt pans within Mbozi Area have apparently been worked for centuries. Remains indicate that by the 13th century people occupying Ivuna cultivated cereals; hunted such wild animals as zebra, bushbuck, warthog, and buffalo; and maintained domesticated cattle, goats, chickens, and dogs. The salt from Ivuna most certainly was a regionally significant resource. However, the role of the salt in trade must remain suggestive rather than positive, since, as Fagan and Yellen (1969:32–33) suggest:

> . . . the absence of imports is an argument that local commodities such as grain and iron tools were handled in preference to foreign or coastal objects. The trading of salt was evidently only part of a complex network of bartering contacts which flourished throughout East and Central Africa during the Iron Age. It is unfortunate that the objects handled in domestic trade do not normally survive in the archaeological record.

Although connections among the Ivuna peoples, occupants of the Kalambo site, and remains found elsewhere in the Corridor would be extremely tenuous (as would relationships to present-day inhabitants), this evidence all points to a very long period of agricultural occupation of the Corridor, certainly longer than Nyiha legends would have us believe.

The salt workings at Ivuna provide one important piece of evidence suggesting a long period of contact and interaction in and around Mbozi from prehistorical time through the present. Never in isolation, the Nyiha, their neighbors, and their forebears have been influenced by their mutual cultural affinities and the literal "march of history" through the Corridor.

The specific origins of the Nyiha are to be found in Bantu migrations peopling the Corridor. While Greenberg had hypothesized a Bantu nucleus in Nigeria and the Cameroons, Guthrie (Huntingford 1963:81) proposed a Bantu homeland in the upper Congo Basin. From here the matrilineal Bantu spread outward after having acquired iron-working techniques. Bantu of the Corridor area came from the Congo Basin west of Lake Tanganyika. Although these people were matrilineal (Figure 2.2), those in the present area of Tanzania and Malawi were later influenced by other cultural groups and became patrilineal. Cattle-holding may have been an important part of this change (Huntingford 1963).

A series of migrations during the 18th century completed the populating of most of the Corridor area. The Maravi group, constituting many peoples of present southern Malawi (Chewa, Nyanja), arrived from the Lake Tanganyika region in the early 16th century and continued their dispersal through the 1700s. A large outward explosion of peoples from the Luba cluster of the Congo Basin began in the 16th century, perhaps carrying improved techniques of iron smelting. The most significant of these peoples was the Bemba who by 1800 arrived in their present tribal area in northwestern Zambia (Wills 1964:48-56).

The last pre-European immigration to the Corridor was that

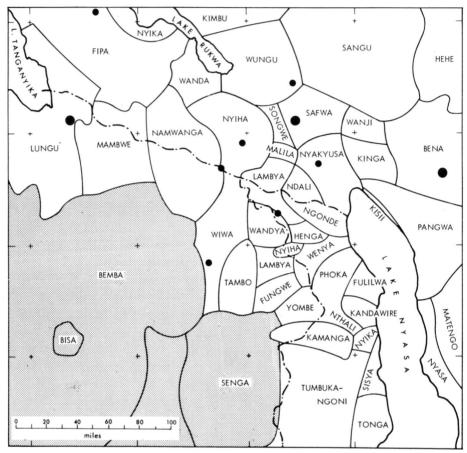

Figure 2.2. Peoples of the Tanganyika–Nyasa Corridor. Areas characterized by a matrilineal pattern of descent are stippled, those characterized by a patrilineal descent are clear. [After Tew (1950); Wilson (1958); Willis (1966); and field reconnaissance (1966–1967).]

of the Ngoni (Barnes 1954). Fleeing from the Zulu, Ngoni arrived on the Ufipa Plateau in 1840, already incorporating refugees of plundered and defeated peoples. In about 1845 the main body of the tribe split. Several groups went south to Malawi, one of them suffering defeat by the Bemba en route. Others raided into the Congo or northeastward toward Lake Victoria. Still another group moved eastward, then south along Lake Nyasa. The Ngoni crossed and recrossed Mbozi. Virtually every Nyiha chiefdom has legends of warfare with them (Brock 1966:3).

    In reaction to the Ngoni, the Sangu united under Mwahawangu. His grandson and successor was the first Merere, the feared Sangu chief who built a fortified settlement at Utengule in Usongwe. He and his successor raised havoc in the Rukwa Songwe valley vicinity for several decades (Oliver 1963b; Wright 1968a).

Having evolved as a local military power during the 19th century, the Sangu under Towelamahamba, the first Merere, formed alliances with neighboring peoples and coastal traders. By the 1890s Sangu influence had contracted and they had been forced to move westward into the present Songwe area. The Nyiha report many instances of defeat and pillage at Sangu hands, and Mwashambwa, at least, came under Sangu rule (Brock 1968:70). Merere sought an alliance with the Germans focused against the Hehe. An arrangement was eventually made for the Sangu to monitor the western Hehe boundary in exchange for restoration of former Sangu territory. Merere's caravans to the coast provided the fastest and most reliable route for German communication. Joint German–Sangu expeditions were launched in the early 1890s, one of which was repelled by the Nyiha chief Nzunda. The successor to Merere (Merere II or Merere Mgandilwa) led the Sangu in the Hehe wars, himself being singled out for a role as paramount chief for western Iringa District. Sangu power was subsequently smothered in the present areas of Usongwe and Usafwa, the Sangu returning to their homeland and the local chieftainship being restored to a Safwa (Wright 1968a).

Copper workings and salt pans in Katanga had attracted Portuguese, Arab, and African traders. Early 19th century trade routes connected the coast with Lake Tanganyika and Kilwa with Lake Nyasa. Arab and Swahili merchants were operating deep in the interior by midcentury. Ivory for European and American markets and slaves for the Arab world and French Indian Ocean islands were the principal commodities (Smith 1963).

The Bemba, Arabs, and their African allies raided widely in the Corridor area for slaves. In addition to their role in the slave trade, it is also likely that the Arabs brought the first continuous contact with foreign goods to Corridor peoples, including cloth and guns. Two major export routes from present Zambia and Malawi converged in Mbeya area and continued to Kilwa and Zanzibar (Gann 1954). Livingstone found Arab slave traders and ivory hunters west of Lake Nyasa in 1863. Slave raiders encountered little resistance from the weakly consolidated peoples of the Tanganyika–Zambia border area. The Bemba, Sangu, and others exchanged captive slaves for arms and ammunition with Arabs who controlled the export routes. In 1881, Stewart (1881:267–268) described the Bemba as "the terror of the country." Boileau (1899:585) saw numbers of men mutilated by the Bemba. The British induced the Sultan of Zanzibar to outlaw exports of slaves in 1873, but through the 1880s, Arab slavers in the northern Lake Nyasa area could not be suppressed. Finally, Arab strength was militarily broken in 1896, and missionary activity prevented any consolidated resistence by the Bemba until they were pacified in 1899 (Gann 1954). However, reports of continued pillage on the German side of the border persisted through 1906 (Willis 1966:xv). Although slaves born after 1905 in Tanganyika were officially free, slavery was only totally abolished in 1922 under the British (Moffett 1958:90–91).

The last century in the Corridor area saw considerable activity culminating in the establishment of the British and German colonial powers by conquest in the 1890s. During this period of activity, salt working at Ivuna and ivory hunting in the Rukwa area were important features of the local economy. The Nyiha and Namwanga were known for their work in weaving, smelting, and iron working; and the Nyiha, in particular, were renowned as hunters and regional traders (Wright 1971:27). This activity among the Nyiha seems to have persisted into the early colonial period, but the Nyiha subsequently withdrew to dependence upon agriculture in a period that seems roughly contemporary with the abandonment of the fortified villages into which they had been forced during many decades of unrest. Perhaps this withdrawal was a natural response to several generations of extremely insecure life and the considerable movement required for survival. During the post-World War II period, however, they have regained economic prominence, now through the development of market-oriented agricultural activity and an impressive process of rural modernization. Before turning to the culture of the Nyiha themselves, it is appropriate to review some characteristics of the peoples among whom they are spatially situated—the peoples of the Corridor region.

Peoples of the Corridor Region

For many of the peoples of the Tanganyika–Nyasa Corridor detailed historical and ethnographic accounts are now available.[2] Most of the Corridor tribes are linguistically related to the Nyiha, being within the same regional zone and, in many cases, the same local group of Bantu languages as Shinyiha (Guthrie 1967:56–57). Several Nyiha neighbors speak languages very closely akin to Shinyiha. In fact, the boundary between languages is not distinct, but gradational. Nyiha find they can easily communicate with Songwe and western Safwa. With eastern Safwa, communication is much more difficult as differences in pronunciation become more marked. Shinyiha is more like Ciwanda than Cinamwanga, which some Nyiha regard as a different language. Nyiha also converse easily with Lambya and Malila. With peoples of Rungwe District (Nyakyusa, Ndali) and Ufipa (Fipa, Mambwe) communications is only possible with use of the Kiswahili *lingua franca*. The Tanzania–Zambia–Malawi international boundary is also the frontier between Kiswahili and Chinyanja as languages of cross-cultural communication. Kiswahili was not widely spoken in Unyiha until after 1930, the region being at the farthest fringes of Kiswahili distribution. Today, most women and some men still have little or no knowledge of Kiswahili, the Tanzanian national language.

The following discussion of Corridor peoples has been organized by linguistic relationships (Figure 2.3) based upon the work of Bryan (1959) and Guthrie (1967, 1969).[3] Areas of

[2]Summaries by Tew (1950), Wilson (1958), and Willis (1966) cover many Corridor peoples.

[3]Corresponding names for the regional Bantu language groups of the Corridor are:

Bryan (1959)	Guthrie (1967, 1969)
Nyiha–Safwa	Nyika–Safwa
Fipa–Mambwe	Fipa–Mambwe
Nyakyusa–Ngonde	Konde
Sukuma	Sukuma–Nyamwezi
Hehe–Bena	Bena–Kinga
Tumbuka–Tonga	Tumbuka; Manda[a]
Bemba	Bemba; Bisa–Lamba[b]

[a]Guthrie places the Tonga within the Manda group.

[b]Guthrie places the Bisa within the Bisa–Lamba group.

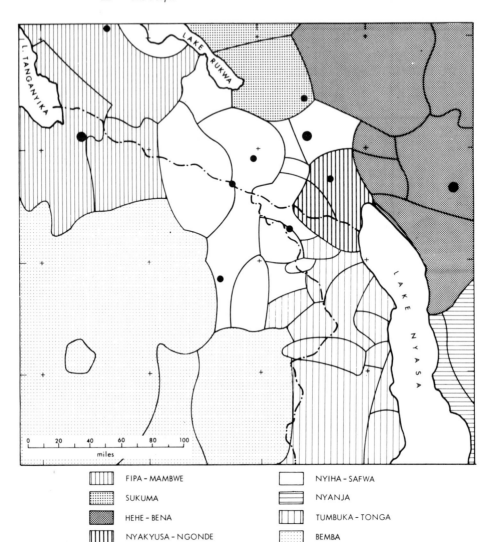

Figure 2.3. Language groups of the Tanganyika–Nyasa Corridor. [After Bryan (1959).]

cultural similarity could also be constructed on other grouping criteria:

1. Common traditions of origin or expressed likenesses
2. Settlement patterns
3. Descent system
4. Indigenous political system
5. Economy

6. Material culture
7. Religious beliefs
8. Rituals (Wilson 1958; Willis 1966:ix)

However, these traits do not always covary, and language provides an arbitrary but uniform method of grouping. Our attention is directed to those peoples (Figure 2.2) and features most relevant to discussion of the Nyiha of Mbozi.

The Nyiha–Safwa Group

The Nyiha–Safwa group includes the Namwanga, Wanda, Songwe, Malila, Lambya, Wiwa, Wandya, Tambo, Nyiha (Malawi), and Nyika (Ufipa). These peoples are all patrilineal, share similar agricultural techniques (Chapter 6), and are historically interrelated. The Nyiha can converse with most of these peoples (in some cases with difficulty) without resorting to a *lingua franca*. Most intertribal marriages of Nyiha are contracted with these peoples, and a number of clan names are common to many or perhaps all of them.

The Songwe. The Songwe are closely related to both Nyiha and Safwa. Their chiefly line traces its history from Nyakyusa or Ngonde people. Three sons of this line who were born in Umalila had small chiefdoms in Usongwe. Sambi ruled over Songwe village while Mwaliyego's village was at Utengule and Mirambo's at Igale. Merere II captured and held Utengule until the Germans gained ascendancy over him. The Songwe chiefs returned and peace was restored. Chief Mwaliyego of Utengule told me that his people really were Safwa, and in many ways the Songwe might be justly considered as a subdivision of the Safwa rather than an independent group. The British recognized one Safwa chief over both the Safwa and Songwe. Today, however, the Songwe have their own local government (Moffett 1958:239). On the basis of linguistic similarities, it is likely that the earliest Nyiha, Songwe, and Safwa were one people upon whom later immigrants imposed chiefly lines and diverse characteristics. The present population of Usongwe consists of people who claim to be native Songwe plus Safwa, Nyiha, and Nyakyusa. Songwe live both in villages and dispersed farmsteads (Mwaliyego 1967).

The Safwa. Wilson (1958) stated that the Safwa were primarily hunters and gatherers to whom cattle and agriculture were late introductions. Harwood (1970) describes the contemporary Safwa as swidden cultivators with livestock and hunting of small game. Safwa chiefs trace their origins to Ukinga, but there is a notable lack of depth in Safwa tradition. They apparently scattered into the mountain areas during the 19th century unrest. Occupation of open lowlands around Mbeya town is a colonial phenomenon. In eastern Usafwa, people claim to understand Wanji and Kinga languages but have great difficulty with Shinyiha (Kootz-Kretschmer 1926:9; Willis 1966:71; Tanzania SHPB). Safwa now live in dispersed homesteads and in villages. In the 1960s, many of their houses were still the round type with conical roofs (Figure 2.4).

The Namwanga and Wiwa. The Namwanga occupy the Msangano Trough and Ufipa Plateau segments of Mbozi, but extend an even greater distance into Zambia. According to Chisholm

Figure 2.4. Safwa traditional house. This variant of the basic regional type has walls of split bamboo. Three elders stand with two *bwana shambas*, agricultural extension officers.

(1910:360), the Wiwa are a "comparatively recent offshoot of
the Winamwanga, and in language, manners, and customs prac-
tically identical with (them)." According to myth, the people
led a hunting and gathering existence in the forests before the
arrival of the great Musyani from Bisa country who brought
iron working and knowledge of agriculture.

   The Namwanga and Wiwa formerly lived in stockaded villages
similar to the Nyiha (Chisholm 1910). One part of the village
wall, *mfoto,* could be easily pushed down from inside as a means
of rapid escape. All features of human habitation were found
within the villages, including the cylindrical wattle and daub
huts, grainstores, pigeon houses, and cattle corrals. The Namwanga and Wiwa today live in both unwalled villages and dispersed homesteads. Many of the houses are now the hip-roof,
rectangular type, but the cylindrical house is much more prevalent than in Unyiha. Houses built in the hotter and drier Msangano area (Figure 2.5) are usually surrounded by an enclosed
veranda for livestock which also helps to cool the house.

Figure 2.5. Namwanga traditional house. At Chitete the basic wattle-
and-daub cylindrical house with conical roof is hidden by the veranda wall
of grass or sorghum stalks. Domesticated pigeons are kept by the Namwanga,
as among the Nyiha.

Land-tenure practices of the Namwanga and Wiwa are similar to the Nyiha (Tanzania MbDB), and these peoples share many clan names. Like the Nyiha, the Namwanga and Wiwa were renowned iron workers, a craft that has recently died out. Their agricultural system is based on the *citemene* and *nkule* techniques (Chapter 6). Cattle, goats, sheep, chickens, and pigeons are raised (Willis 1966:32-39). The Namwanga of Msangano have seen development in Unyiha and sharply feel the consequence of their isolation. In 1967 they were less than optimistic for any major road improvement. They have concentrated on cattle production in recent years; the Msangano Trough seems to be free of both sleeping sickness and East Coast Fever.

The Wanda and Wandya. The Wanda occupy the northern portion of Mbozi Area (Kamsamba Division) in the Rukwa Valley. Wanda and Nyiha languages are similar but not identical. Wanda traditions relate a chiefly origin in Unyiha, with their chiefdom located at Kamsamba, and a minor chiefdom at Ivuna. A number of clan names are common with the Nyiha (Willis, 1966:60–61). The Wanda live in concentrated settlements and cultivate finger millet, sorghum, bulrush millet, and maize as staple crops. Like the Nyiha the round house was traditional but has been replaced by the rectangular type. As among Namwanga of Msangano, there is a focus on cattle production for sale at monthly cattle markets.

The Wandya of Zambia claim to have broken off from the Wungu, and moving south defeated a Kamanga enclave whose survivors in turn moved south. It is not impossible that the Wandya were related to the Wanda of Tanzania. The Wandya were frequently raided by the Bemba (Brelsford 1956).

Within a small enclave in the Wanda chiefdom (Mkulwe) live the Kuulwe. These people trace their origin from Zambia and Malawi, supposedly arriving in Uwanda *ca.* 1860–1870 as a result of Ngoni raids. They live in a dense cluster of concentrated villages and have politically been absorbed by the Wanda. They share a number of clan names with the Nyiha and other tribes. Their subsistence patterns seem to be similar to the Wanda (Willis 1966:65–66).

The Lambya. The Lambya are closely related to the Nyiha with whom they frequently intermarry. These peoples have a number of similar clan names. Like the Nyiha they lived in stockaded villages in the 19th century (Kerr-Cross 1890:286; Fotheringham 1891:129–33). Most of the Lambya in Tanzania live in Rungwe Area, but a small number live directly across the international boundary in Zambia and Malawi. A detached group lives further south, adjacent to the Tambo and Fungwe. According to Brelsford (1956), these are one people.

The Malila. The Malila are related to the Nyiha, their languages being virtually identical. Malila claim to be able to converse much more easily with Nyiha than Safwa or Songwe. They live in the hills just east of Mbozi Area. Their agricultural practices are very similar to those of Iyula Division in Unyiha.

The Tambo. The Tambo were a group that broke away from the Bisa as a result of population pressure, settling in the area adjacent to the Wiwa, Lambya, and others. They are most closely related to the Wiwa (Brelsford 1956:76).

The Malawi Nyiha. The Malawi Nyiha are an offshoot of the Mbozi Nyiha. In the course of migrating south after dissension in Mbozi, they helped the Fungwe defend themselves against the Ngoni. In turn, they were given an area in which to settle (Brelsford 1956:78). One Mbozi Nyiha informant had visited the Malawi Nyiha and returned with a wife. He reported that the languages were nearly identical with the exception of nouns relating to crops and material items which correspond to the terminology of the local region.

The Ufipa Nyika. The Ufipa Nyika are also related to the Mbozi Nyiha (Willis 1966:68). They probably separated from the main body of the Nyiha during Ngoni raids (Popplewell 1936:100). They settled on the Ufipa Plateau, but took refuge in the hills forming the escarpment to Lake Rukwa after having been attacked by the Fipa. The Ufipa Nyika practice farming techniques identical to the Fipa (Willis 1966:68–70).

The Fipa. The Fipa occupy the treeless plateau as well as portions of the Lake Tanganyika shore and Rukwa Valley. They live in widely spaced villages located near perennial water supplies. The Fipa trace their origin to northeastern Zambia, where according to myth the dominant political structure was founded by light-skinned women who came from the north and married Nyiha hunters (Willis 1966:18–19). Fipa are members of kin groups called *uluko* (Willis 1966:21–22), similar to the Nyiha *uluho* or clan.

The traditional house style of the Fipa was the cylindrical style similar to the Namwanga type illustrated in Figure 2.5. The rectangular wattle and daub hut is now ubiquitous. Like the Nyiha, the Fipa and Mambwe were skilled iron workers, but among the Fipa the tradition actively persists (Fromm 1912; Greig 1937; Willis 1966, 1968).

The Mambwe. According to Watson (1958), the Mambwe arrived in their present area before arrival of the Bemba. The chiefly line from the Congo arrived to find people who already practiced agriculture. They suffered at the hands of the Ngoni and Bemba, and lived in large palisaded villages akin to the Nyiha (von Wissman 1891: 270–271). Now, they live in smaller villages. Some legends attribute chiefly lines to a western origin, while two subchiefly lines are partially related to the Nyiha. Some clan names are common with the Nyiha.

Northern Mambwe who occupy the southern Ufipa Plateau practice agricultural techniques like those of the Fipa. In the southern *miombo (Brachystegia)* woodland area, the Mambwe make *citemene* fields as well. Both the Fipa and Mambwe keep large numbers of cattle. Among the Mambwe, rectangular houses have replaced round house style.

The Lungu. The Lungu live in concentrated villages and practice the *citemene* agricultural system. They originated from the west and suffered greatly under the Ngoni, Arabs, and Bemba. Many Lungu clan names are shared with the Mambwe, Namwanga, Wiwa, and Nyiha (Willis 1966: 40–46).

The Nyakyusa–Ngonde Group

The Nyakyusa and Ndali of Tanzania and the Ngonde of Malawi are closely related and speak slightly different dialects of the same language. The age village which forms such a distinct cultural feature among the Nyakyusa and Ngonde is not found among the Ndali (Wilson 1958). Historically, the groups differ in that the Ngonde had a centralized chieftainship related to to the development of trade for ivory, while the Nyakyusa, like the Nyiha, maintained many small independent chiefdoms until European penetration. Tradition suggests that both the Nyakyusa and Ngonde chiefs came from Ukinga and established hegemony over peoples of north Nyasa. Nyakyusa have absorbed several smaller groups and have expanded from their densely populated homeland by migration to both the north (Usafwa) and west (Umalila, Ulambya, and Unyiha). Ngonde people trace linkages to the Namwanga and Wiwa peoples to the west. Nevertheless, the Ngonde–Nyakyusa peoples are markedly distinct from their neighbors in language and culture (Wilson 1963:1–5). Nyiha claim Nyakyusa are identifiable by facial features.

In recent times many Nyakyusa and Ndali have migrated to local and international labor markets. Large numbers have come to Mbozi since the 1930s to work on European coffee estates. There, Nyakyusa are still regarded by European managers as better and more reliable workers than most others. On at least one estate, verbal work orders are issued by the manager in the Kinyakyusa language. In other areas to which they have migrated, Nyakyusa are also considered aggressive, reliable, and able workers. These may be perceived rather than real differences, although migrants are unlikely to be a random sample of any population. In addition, local laborers may make their own farms a primary concern even when employed elsewhere, suggesting that as emigrants obtain local farms any real differences will evaporate.

Other Groups

The Sukuma Language Group. This group, which extends north to Lake Victoria, is represented among the Nyiha neighbors by the Wungu of Chunya area. In spite of the shared boundary with the Nyiha, there is little interaction between these people. They are separated by unpopulated hills forming the escarpment to the Rukwa Valley plus the dry valley itself. The origin of the Wungu seems to be from the north (Ukimbu), although the chief's title is *mwene* like that of the Nyiha, Safwa, and others. They live in concentrated settlements (Willis 1966: 70–71) and grow maize and finger millet as staples. There is a tradition that the Wungu made executions on the steep precipice of

Kwimbi Mountain on the southeastern edge of the Rukwa Valley (Ellis 1957). Sacrifices were made of pure white oxen and goats, supplied by the Nyiha and Namwanga as late as the middle 1930s (Rushby 1965:127–28).

The Hehe–Bena Language Group. In this group is included the Kinga, a people to whom the Nyakyusa and other Corridor groups attribute some chiefly relationship. They live in the high mountain country north of Lake Nyasa. The main crops cultivated are wheat, peas, potatoes, oats, and barley. Wanji are related to both the Kinga and Bena. They have made important strides in planting pyrethrum in recent years in addition to the crops listed above. A few Kinga and Wanji have come to Mbozi as laborers, but Nyiha agriculture has not been altered as a result of contact with them, since Ukinga is much too cold and moist to be considered analogous to Mbozi. The Sangu have been primarily pastoral, but a colony of Baluchi settlers in the Usangu flats has introduced irrigated rice cultivation which may transform the Sangu economy as well (Moffett 1958:240). Sangu military interaction with the Nyiha was significant until the removal of Merere from Utengule; today, the rice farms have attracted a few Nyiha labor migrants. The Hehe, Bena, Pangwa, and Kisii, as well as the Nyanja-speakers of Southern Tanzania, are considerably more removed from Unyiha and will not be discussed here.

The Tumbuka–Tonga Language Group. A number of peoples of northern Malawi and adjoining Zambia, as well as the more numerous Tumbuka and Tonga, are included in this group. The Tumbuka and Kamanga are of some antiquity in the Corridor area. It is possible they were part of the Maravi migrations. All of the peoples of this cluster were traditionally matrilineal, but this pattern was altered to patrilineality with the arrival of the Chikuramayembe or Ngoni. The Tumbuka, Kamanga, and Henga languages are dialectal forms of one language, and their recent history is closely related. In the late 18th century a group of traders assumed local chieftainships over the disunited, dispersed Kamanga and Tumbuka villages. The Henga were created as a separate entity at that time, and the Sisya, Fulilwa (migrants from the eastern side of Lake Nyasa), and Kandawire (traders from the south) were under the·

Chikuramayembe control. The Ngoni overthrew the Kamanga dynasty and subjected the Tumbuka to serfdom. Tumbuka customs were revived after the British subdued the Ngoni in 1896. Phoka, Yombe, Fungwe, and Wenya languages are also Tumbuka dialects (Tew 1950). However, Brelsford (1956) has noted close relationships of Yombe and Fungwe chiefs to the Ngonde. There are dozens of lesser peoples, the Lake Nyasa Nyika and Nthali among them, whose relationships within the cluster are uncertain. Mound-cultivated maize is the staple crop for most of the Tumbuka tribes.

The Tonga, like the Tumbuka, are probably part of the Maravi peoples. The Tonga language grades through the Sisya into Tumbuka as a continuum. The Tonga and Sisya are primarily fishing people who also cultivate cassava, rice, maize, potatoes, and beans as staples (Tew 1950).

The Bemba Language Group. This language group is represented by the Bemba and Bisa. The Bisa claim to be part of an earlier migration from the upper Congo basin, while the Bemba assert that the Bisa broke away from them during the migration process. According to Trapnell (1943), the Bisa arrived in the 17th century. In the early 18th century, the Bemba pushed the Bisa south, leaving a small enclave within the Bemba area. Bemba agricultural practices, according to Trapnell, were learned from the indigenous Mambwe, Namwanga, Wiwa, and other peoples pushed aside by the Bemba expansion. The Bemba state was consolidated in the middle 19th century under the impetus of Arab trade of guns for slaves and ivory. They ruled or harassed much of the western Corridor area until subdued by the British. The Bemba and Bisa speak nearly identical languages (Brelsford 1956:20–37).

The Nyiha of Mbozi are part of larger culture groups. On the broadest scale, they are Bantu-speakers and represent one segment among the many Bantu migrations. More specifically, they are historically and culturally interrelated with many of the peoples of the Tanganyika–Nyasa Corridor Area, who in turn have many relationships extending beyond this immediate region. Finally, the Nyiha are part of a cluster of closely related peoples sharing very similar languages. Hence, traditional Nyiha culture (and agriculture) has not evolved in isolation but affected and was affected by other peoples and ideas. In recent decades, the pace of change has hastened as the Nyiha have been exposed to totally alien cultures and ideas whose eventual impact *may* be greater than all their previous experiences.

The Nyiha[4]

At the time of European contact, the Nyiha had 12 small chiefdoms whose traditional seats were in the following modern Village Development Committees (see front end papers):

1. Nzunda: Vwawa
2. Mwamengo: Mlangali
3. Mugaya: Songwe (Nsenjele)
4. Msangawale: Itaka
5. Shombe: Ilindi (Iyula)
6. Mwasenga: Iyula–Hezya
7. Nzowa: Wanishe–Igamba
8. Mwamlima: Iyula
9. Mwangamba: Hanseketwa
10. Mwashambwa: Ndolezi (Mlangali)
11. Mwezimpya: Ruanda
12. Mwalembe (Umwembe):   Ihanda

Of these chiefdoms, Mwezimpya seems to have been created by a fission of the Mwashambwa chiefdom; Mugaya may have emerged since German colonization (his name is absent from their lists or maps); and Mwalembe's precolonial status is uncertain, the chief having been ousted by Nzowa in the early colonial period (Brock 1968:62). Accounts of the origins of the chiefly lines justify their rule through assertions of "first occupancy of unoccupied land, and claims to have introduced superior technology and social organization to a more primitive people [Brock 1968:63]." Nzowa and Mwamlima clans trace their origin to the north and claim to have found a very primitive hunting

[4]Social anthropologist Beverley Brock lived among the Nyiha for 6 months in 1961. Portions of the discussion of Nyiha culture are derived from her work (Brock 1963, 1966, 1967–1973, 1968, 1969) as corroborated in the field in 1966–1967. Mariam Slater spent a year (1962–1963) carrying out anthropological research among the Nyiha (Slater 1966). She has a monograph in preparation.

and gathering population when they arrived (Mwamlima 1967). Chief Mwamlima said his ancestors were related to chiefly lines among the Sangu, Safwa, Gogo, and Malila. The first six chiefs are part of a common line (the Simbeya group, after their common clan name) who came from Zambia as traveling groups who claim to have occupied uninhabited lands. Chief Mwasenga (1967) said his ancestors were Wiwa people who came from the area now occupied by the Bisa. The remaining lines are also from the south and southwest.

Oral legends relating to Nyiha history have been interpreted by Brock (1968) and can be summarized briefly here. References applying to the period before the mid–19th century account for establishment of the chiefly lineages at an uncertain date and report disagreements among the petty chiefdoms with flexible alliances that were later broken, former allies fighting among themselves. The long era of raiding by Ngoni, Sangu, and Bemba as well as by the Namwanga and Wungu is recollected. In

Figure 2.6 Construction of a grainstore. The Nyiha *ishanga* is made from split grass *(amatete)* tied where necessary with rope made from the cambium layer of trees *(inyezya)* but otherwise woven. The store will be set on a small platform 1 or 2 feet above the ground. A removable, conical thatched roof *(ishisondje)* will be added.

response to this incessant threat of death, the Nyiha and neighbors moved considerably to escape raids; and although they fought bravely, they were never cemented into a unified political organization. They maintained the petty chiefdom system through the colonial imposition. Indeed, the Nyiha met early German incursion with the same tenacity and resistance they had exhibited against the Sangu.

Legendary immigrants from the southwest brought hoes to cultivate traditional finger millet, sorghum, and other crops. Spears *(impalala)*, hoes *(ijembe)*, axes *(imbinzo, intemo)*, knives *(ifwu, ishisu)*, and machetes *(insengo, ishipanga)* were made by specialized iron workers *(umusyani)* from *mbuga* bog iron *(inyimbo)* smelted in tall, cylindrical furnaces *(ilingu;* B. Brock and P. Brock 1965). The Nkota clan of Wiwa origin has been

Figure 2.7. Making a basket. Men make several varieties of baskets *(ishitundu)*. This is one of a general type for carrying harvested grain from the field. A shallower type is used for winnowing grain.

important among master smelters. The Nkotas are related
through the Wiwa to the Namwanga, whose founder, Msiani
(Musyani), according to legend, was the son of an immigrant
to Bisa country from the Luba Kingdom of the upper Congo
Basin (B. Brock and P. Brock 1965). Iron smelting succumbed
to competition from mass-produced imports early in this century,
but scrap iron is still forged into spears, axes, and knives. The
leaves of an automobile or truck spring are considered ideal
for this purpose.

There was a specialized group of elephant *(inzoru)* hunters
who used heavy spears produced by the master ironworkers.
Ivory was traded with the Arabs as well as slaves (Brock 1963:55).
There are still a number of elephants in southern Rukwa Valley
who roam the uninhabited hills that bound the Mbozi Plateau
on the north. Elephants occasionally menace agriculture in Nambinzo as do vervet monkeys *(imbadji)* and yellow baboons
(intumbi). Today only small game such as the mongoose *(usinde)*,
hare, squirrel, mice *(imbera)*, and mole *(utunko)* populate settled
areas of Unyiha. These, as well as birds, are trapped and eaten.
Monkeys are a pest in the eastern hills, but Nyiha have a strong
aversion to harming them. Occasional reports of a leopard seen
at night were heard in 1967, but it is unlikely such a large cat
would be unnoticed during daylight hours. There may be, however, a small number of the East African civet cats *(umutolo)*
which in panic could be identified as leopards.[5]

Men are the hunters *(umulumba)* among the Nyiha; they also
clear new fields; help in their cultivation; have general charge
of livestock tended by young boys; and make baskets, grainstores, mats, and tools (Figures 2.6, 2.7, 2.8, and 2.9). They
also place hollow logs in trees for beehives and process the
honey. Women help in field preparation; sow, weed, harvest,
and thresh the crop; prepare food; and make clay pots. The
former scanty clothing of barkcloth *(imfumpa)*, or woven wild
cotton *(amatote;* Kerr-Cross 1890:289) has been completely
replaced by western dress. Recently, men have taken almost

[5]As early as the turn of the century, large game was found only in the Rukwa
Valley (Great Britain Admiralty 1916:260). It is here that large remaining populations of impala, zebra, buffalo, hippopotamus, elephant, waterbuck, bushbuck,
and giraffe are found. Poaching has been all but eliminated, and there is some
hope that a game park might be created in the Rukwa Valley, although it would
be outside of Mbozi Area (Rushby 1965).

Figure 2.8. Braiding a grass mat. The mat is formed by sewing sections cut from a continously braided strip. This type of mat *(ishilili)* may be used for sleeping or as a carpet. A mat woven from soft reeds is preferred for sleeping over the type shown.

Figure 2.9. Carving an ax handle *(ishipinyi)*. The carving adze *(imbinzo)* has the blade set perpendicular to the shaft, while the ax used for felling trees *(intemo)* has the blade set parallel to the shaft.

full responsibility for coffee production, with help from their wives. This may, in fact, be a redirection of time and effort no longer devoted to hunting and defense.

Kinship, Marriage, and Inheritance

Nyiha are organized in dispersed exogamous clans known as *uluho*. Large clans become subdivided into smaller groups taking the name of an important ancestor while recognizing membership within the larger group. Long-separated subclans may eventually relax the rule of exogamy, applying it only to the smaller group (Brock 1966:7). Inheritance and succession are usually controlled by lineages within the clan, also known as *uluho* (Brock 1968:63). While the clan name often serves as a surname, I found that men frequently give their father's name as a surname, and one must ask directly to determine *uluho*. Married women continue to use the *uluho* of their father, while descent of the children is traced through the father. Polygyny was and is common.

Marriage is traditionally a lengthy and expensive process. The man is required to provide both bride service (labor for the bride's family) and bride wealth. Traditional amounts were several years of service, beads, cloth, a few goats and sheep, and less commonly a cow. Brock (1966:9) reported that in 1961 bride service was much decreased or replaced by a greater bride-price. Typical bride-price at that time had grown to "8 to 10 cows, plus sheep and goats, hoes, blankets, and 100 to 300 shillings in cash." In 1967, the bride-price still required goods worth 3000–3500 shillings, but several young people interviewed talked of eloping to avoid this financial burden. There is an obvious interrelationship between economic growth and rising bride-price. Higher amounts are at once an impetus to and a result of development.

On death, the wives of a married man are inherited by a brother or son with preferences on both sides taken into account. A son might have his own mother live with him, but not inherit her as a wife. Traditionally, a man's wealth and duties passed to his brother, but today these go to his eldest son, as do his rights to land.

The Supernatural

Nyiha traditional religion included a god, *Mulungu;* spirits of unknown origin; and ancestors to whom prayer and sacrifices were offered before most activities were begun (Figure 2.10). When disaster strikes a person or family, a diviner is necessary to diagnose it. Matters that concern the whole society, however, require the chief's leadership. Two regular rituals were held each year, in November before sowing and in May before the harvest. All villagers gathered at the chief's home, and prayers were made to the chief's ancestors for success in the agricultural efforts to come. The chief then made a new fire, and from this each family rekindled the fire in its home (Bachmann 1943:129–132; Brock 1966:19–21).

Some of the older chiefs still make the new fire at planting time, but the practice has, in general, disappeared. One group of elders explained decreasing crop yields in recent years as a result of neglect of the new fire ritual. In the case of rain failure or a disease epidemic, the chief traditionally prayed and

Figure 2.10. The offering hut *(ahasana).* This small feature of the cultural landscape marks the grave of a chief or respected elder. Offerings *(imfindjile)* of chicken or beer are made here to assure protection against danger to the family.

made sacrifices at the secret burial place of his ancestors (Brock 1966:21). Whether this practice still continues is uncertain.

Nyiha have great confidence in medicine. While some medicines are used for ritual or magical effect, Nyiha also concoct a number of folk remedies using herbs, roots, bark, and leaves for application or ingestion. Their affinity for medicine has affected acceptance of dispensary services. In fact, people in Nambinzo VDC built a self-help dispensary that still had not been staffed by the government several years later. Brock (1966) observed that belief in the ability of ancestors to punish those failing to maintain kinship obligations has been decreased by a combination of individualism, education, and Christianity. However, she has suggested that accusations of sorcery may have increased during the process of modernization, a hypothesis demanding attention in later research (Brock, personal communication). In the realm of disputes over land alone (Chapter 8), it would appear that tensions could be sufficiently great to lead to these accusations.

Most of the Christian Nyiha follow the Protestant Moravian or Roman Catholic faiths. The African National Church has made some inroads since it freely allows multiple marriages. Although once tolerant of polygyny (Wright 1971:106), Moravians are now strict on this issue, as are the Roman Catholics. However, more than one White Father expressed a personal view that prohibition of multiple marriages obviously did not originate from the mission field nor exhibit an understanding of inheritance and care of widows, sexual prohibitions, and the like. Until these prohibitions do recognize local traditions, the doors of the church should not be closed to polygynists.

The Political System

In the late 19th century, the Nyiha were divided among the 12 minor chiefdoms. Each was headed by a chief *(mwene)* under whom were one to three councillors *(awahombe)*; an amorphous group of elders *(avasongo wa nunsi)*; the chief's kinsmen *(wanamfumu)*; and two major court officials, the *mugave* or the chief's bodyguard, and the *mutwale*, the head messenger and carrier of the stool and other symbols of the chief's office. The individual chiefdom was a territorial unit with clearly defined boundaries. Membership in the chiefdom was based on residence, and citizens could change their allegiance by moving to another chiefdom (Brock 1966:14):

> The chief's store-houses of grain were a form of famine insurance for his people. He was also expected to provide animals for ritual sacrifice on behalf of his chiefdom. Thus the chief had larger fields, more wives, and more food than his subjects, but he also had more demands on these resources. The rights of tribute from a small population were divided among eleven or twelve petty chiefs and this very division must have reinforced the norm of generosity, for those dissatisfied with their chief could move to the country of one considered more liberal.

Internally, the chiefdom consisted of several palisaded villages each having a headman. This settlement pattern was ubiquitous in Unyiha during the late 19th century and was associated with the unrest of that period. While temporary and unstable alliances were occasionally formed by two or more chiefdoms, there was never a paramount chief.

Palisaded villages ended with the cessation of raiding. The predominant settlement pattern is now dispersed homesteads separated by cultivated fields and bush. Some areas retain village-like status by a clustering of homesteads, but the territorial unit sometimes called a village now is really a neighborhood having imprecise boundaries. In the early 1960s several of these neighborhoods were under one headman (Brock 1966).

Mbeya District was established by the British colonial government in 1926. Administration there as elsewhere in Tanganyika was of the indirect variety—ruling through the traditional political structure (Perham 1931). Native administrative units were legally recognized by the government and controlled in the main by advice and supervision. In order to implement indirect rule, the government hastily collected information about cultural institutions, much of which was recorded in the district books. This information was often incomplete and uneven in quality as well as quantity. Sir Donald Cameron, Governor from 1925 to 1931, brought indirect rule to Tanganyika from his previous post in Northern Nigeria. Recognizing the potential benefits from amalgamation of native authorities to form larger governing bodies, Cameron encouraged this evolution from indirect rule when local peoples felt they would gain from formation of larger units (Ingham 1965:575). The government thus attempted to amalgamate the Nyiha with the Namwanga and Wanda under the Namwanga chief, Mkoma, in 1928. This was found to be unworkable because of the lack of close relationship between the peoples involved and the resentment of other chiefs to being subordinate to Mkoma. Although treasuries remained joint, the association was dissolved in 1929 (Tanganyika ARPC 1929:54), suggesting that national policy rather than local support for unification had been the impetus for amalgamation here.

The British then attempted to institute a paramount chief among the Nyiha. This failed as did a ruling council consisting of all 12 chiefs. A tripartite division of Unyiha (the present Iyula, Vwawa, and Igamba) was finally settled upon with each subdivision ruled by a government–recognized chief who presided over a local court. Under each chief served headmen, salaried tax collectors who covered areas occupied by up to 500 families (Brock 1966). These areas were not functionally tied by kinship bonds or patterns of social interaction in marked contrast to the postindependence development of local government. At independence, Mbozi subdistrict (of Mbeya District) consisted of three local courts in Unyiha plus two courts in the Namwanga chiefdom and a court at Kamsamba near Lake Rukwa which was subsidiary to the Namwanga chief. The small settler community had failed in its attempt to secure separate district status for Mbozi before independence. However, Mbozi Area was finally created as a separate administrative unit in 1964.

The present structure of local government in Mbozi has evolved since independence. The roots of the Village Development Committee (VDC) system appear to have been in Mbeya Region, and the idea has become a major facet of national development policy and political organization (Yeager 1968). Recently, to the VDC system has been added the *Kumi-kumi* or Ten–House group—the smallest unit of political organization (Figure 2.11).

The local government structure represents a hierarchy linking the individual through his household to the central government in Dar es Salaam. In addition, the same structure links the individual to the Tanzania African National Union (TANU—the national political party) hierarchy. Not included in the diagram

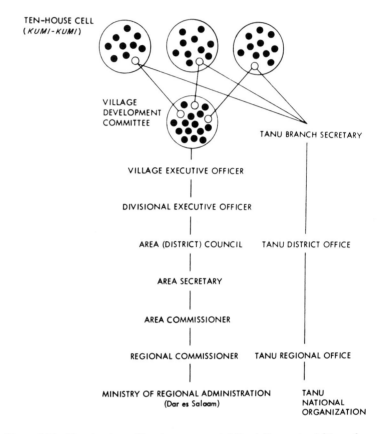

Figure 2.11. The structure of local government, Mbozi, Tanzania. A hierarchy links the individual household to the national government.

is the linkage, more direct but more impersonal, of the elected representative to the National Assembly.

Every 10 (sometimes 11 or more) homesteads are organized in Mbozi into a *Kumi-kumi* group. These houses are usually portions of one neighborhood and are a cluster defined mainly by proximity. The heads of the households elect one among themselves to be head of the *Kumi-kumi* and its representative on the Village Development Committee. The head also serves as an arbitrator of local disputes and representative of and to TANU under the TANU Branch Secretary.

The Village Development Committee is both a body composed of the heads of the *Kumi-kumi* cells and at the same time a well-defined locale in spatial nomenclature.[6] Ideally, the Village Executive Officer is elected from the members of the Village Development Committee. This was the case in Mbozi in 1966, but in 1967 the Village Executive Officers were rotated to new VDCs at the direction of the Divisional Executive Officer. The Village Executive Officer, a full-time salaried official paid by the District Council, is the major tax-collecting agent and acts as a minor arbitrator as well as chairman of VDC meetings and overseer of self-help projects. In Mbozi his duties are closely analogous to those of the former headman. In most cases the VDC territory is identical to the area of a former headman's territory, but the Village Development Committee *creates* a pattern of social interaction that gives unity to its areal extent.

The Divisional Executive Officer oversees the Village Executive Officers for some 5 to 12 VDCs in Mbozi. There are six of these Divisional Executive Officers and 44 VDCs, including those in Unamwanga and Uwanda. The District Council is an elected body, but in 1967 the first Mbozi District Council was yet to be formed. The office with that title in Mbozi was really an extension of the Mbeya District Council in which Mbozi was represented as part of Mbeya District until 1964. Linkages then extend through the Area Secretary and Area Commissioner to the Regional Office of the Ministry of Regional Administration and thence to Dar es Salaam.

Education

Moravian missionaries established primary schools in Mbozi before World War I. By 1929 they operated some 20 schools in Unyiha (Rungwe Mission Archives). It was not until 1936 that these schools were of sufficient quality to receive government aid (Brock 1966:24). Virtually all of the schools were reorganized with government staffing after World War II. The number of primary schools continues to increase. Thirty-six schools offered at least grades one to four in 1967; an additional

---

[6]In the period since 1967, the Village Development Committees in Mbozi have been agglomerated as Ward Development Committees (Barker 1973). For purposes of our analysis here, reference to the political structure contemporary with field research is maintained.

14 locally supported TAPA schools (Tanganyika African Parents Association) offered grades one and two. A number of unofficial local schools with a part-time teacher hired by parents also existed. The Mbozi Upper Primary School opened in 1954 (at the Mission) and was followed by the Mlowo Seminary at the Roman Catholic Mission in 1956 and by Vwawa Upper Primary School in 1958, all of them offering grades five through eight. Students seeking secondary education have to attend school in Mbeya or more distant points.

Education was not compulsory in the 1960s in Tanzania. If a school is available, parents are frequently torn between sending their sons to school or keeping them home to tend cattle. Usually a boy will enroll in school as soon as a younger brother or cousin·can take over his herding responsibilities. Students who are sufficiently fortunate to finish grade five face a qualifying examination to see who continues on to upper primary school. Examinations are also taken at the end of grades eight and twelve to determine which students will attend secondary school and university. For those who do qualify, secondary school and university education is government supported. Others may attend a private secondary school and university if able to afford it. Only a very small percentage of those who enter primary school finish secondary school, and an even smaller number attend university. Brock (1966:24) stated:

> . . . it is surprising how many Nyiha men there are who have had more than primary education—obtained at much sacrifice by parents who persevered and sent their sons elsewhere to schools. Since all schools involve payment of fees, the spread of education, slow as it has been, has been possible only because of the concomitant development of sources of cash income, plus a willingness of fathers to forego the labor of young boys at their usual tasks of goat and cattle herding.

Indeed, provision of school fees for one's children is itself a significant impetus to generation of a cash income. Formerly, education was offered only to boys but enrollment of girls is now increasing in the lower primary schools, while the upper primary schools remain dominantly male. An enlightened educational policy could change the balance of education toward greater sexual equity.

Population Patterns

Distribution of settlement in Mbozi (Figure 2.12) clearly reflects avoidance of areas in which water supplies are ephemeral, eroded edges of the Mbozi Plateau and dry locales in the troughs and Rukwa Basin.[7] The Ufipa Plateau (Unamwanga) is also unevenly populated, reflecting distribution of water supplies in the well-drained portions, and the vast *mbuga* areas in the west (see back end papers).

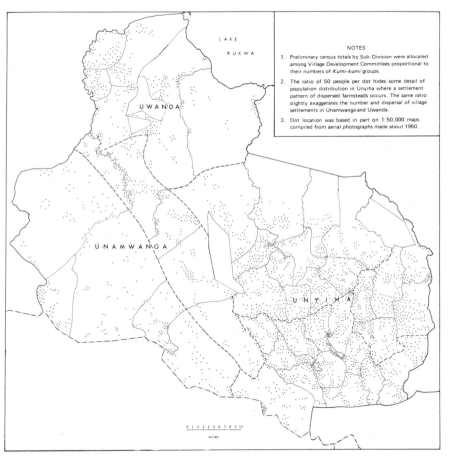

Figure 2.12. Population of Mbozi Area, 1967. One dot equals 50 people. [Source: 1967 Population Census of Tanzania.]

[7]Gillman (1936) emphasized the strong correlation between population distribution and availability of domestic water in Tanganyika.

Population data of varying reliability are available for Mbozi Area from 1900 (Tables 2.1 and 2.2). Differential rates of population growth are due to early introduction of medical services in Unyiha as well as immigration to that chiefdom by other tribes seeking job opportunities and land. Greater prosperity of the Nyiha has probably meant slightly longer life expectancies, although the 1967 census indicates a persistence of high infant mortality (Thomas 1972). The Mbozi Mission dispensary and, later, the government dispensary in Vwawa have made important contributions to the health of those to whom they are accessible.

Within Unyiha, as population has grown, an outward movement from the Nyiha core in the central Mbozi Plateau has taken place. Many men over 50 living in the study areas in northwest-

Table 2.1

Mbozi Area Population, 1900–1967

Year	Total African population by Chiefdom[a]		
	Unyiha	Unamwanga	Uwanda
1900[b]	6–7000	———— 1500 ————	
1916[c]	6,422	5,630	
1926[d]	12,000	4,500	3,000
1928[e]	17,315	10,071	4,214
1930[f]	18,000		
1931[g]	25,179[h]	10,147	5,754
1948[i]	49,754	16,866	8,112
1957[j]	66,285	20,607	10,756
1967[k]	102,263	27,890	17,336

[a]Data for the years 1928, 1931, 1948, 1957, and 1967 are from official censuses; all others are estimates.
[b]Fülleborn 1906:488.
[c]Great Britain Admiralty 1916; data for Unamwanga probably includes Uwanda.
[d]Gillman 1927; Tanzania. Iringa Provincial Book.
[e]Tanzania. Mbeya District Book.
[f] Bagshawe 1930.
[g]Tanzania. Southern Highlands Provincial Book.
[h]Includes the Nyika of Ufipa.
[i]East African Statistical Department 1950.
[j]East African Statistical Department 1958.
[k]Thomas 1968a, b.

Table 2.2

Population Census, 1967[a]

Locale	Population	Area (sq. miles)	Population per sq. mile	Acres per person
Igamba	40,118	841	47.7	13.4
Iyula	19,635	296	66.3	9.7
Vwawa	42,510	513	82.9	7.7
Unyiha	102,263	1,650	60.8	10.5
Uwanda (Kamsamba)	17,336	716	24.2	26.4
Msangano	14,366	445	32.3	19.8
Ndalambo	13,524	889	15.2	42.1
Unamwanga	27,890	1,334	20.5	31.2
Mbozi Area	147,489	3,700	39.9	16.0

[a]Data are from Thomas 1968a, b.

ern Nambinzo and western Wasa VDCs were born in more central locations. There seems to have been a general movement, initiated probably during the German period, into hilly areas on the margin of Mbozi Plateau. This was undoubtedly associated with a search for new areas to clear for agriculture. The frontier status of these marginal VDCs is discussed in later chapters. After the middle 1930s the impetus to this outward expansion was slowed as coffee production became common, sales of surplus traditional crops to Asians became possible, and jobs on coffee estates were plentiful. It is general known that the west and northwest are less suitable for coffee cultivation because of decreasing rainfall and less fertile soils. Hence, a further differential population change has occurred within Unyiha, with greatest population densities occurring in central Mbozi. Densities in the upland areas of Igamba Division are not significantly different than those of Iyula or Vwawa, but are numerically altered by the virtually empty areas in the north and northwest.

The following compound annual population growth rates have occurred in Mbozi in the last decade, as calculated from population census figures for 1957 and 1967:

Unyiha   4.3%
Unamwanga   3.1%
Uwanda   4.9%

It has been impossible to assess the role of immigration in these growth figures. Continuous arrival of Nyakyusa and Ndali into central Mbozi has occurred. There also exists the possibility that the 1967 census was more complete than its 1957 predecessor and that enumeration area boundaries were not identical. However, it is likely that an annual natural growth rate, exclusive of immigration, of about 3% applies. The impact of population growth is evident in more densely settled areas of Unyiha where decreasing amounts of land available per person and the allocation of land to permanent coffee and pine trees have induced a marked alteration of traditional food-crop agriculture as well as land tenure.

Given the directions of present agricultural evolution, population density does not appear to be reaching any critical proportion in Mbozi. Many young people interviewed are less keen on having multiple wives than their elders. Many are concerned about the problems of supporting a large family. Such concern has yet to be manifest in average family size, but it is an indication that should it be desired to institute population control measures, such ideas might be willingly received. It is likely that economic and other motives will affect population growth in Mbozi before problems of "overpopulation" will be evident.

Interestingly, traditional Nyiha prohibitions against sexual relations between a man and a nursing wife seem, on very limited evidence, to be respected. The Nyiha explanation is that if the woman gets pregnant too soon, the nursing baby will die. This seems logical, because a baby weaned too early may have a much lower probability of surviving the crisis of a low-protein diet. This prohibition may in part explain the appeal of polygyny, and certainly functions as a control on birth rates.

The Plural Society

From the late 1920s, the population of Mbozi has assumed plural characteristics. Before World War II, 63 estates were occupied by European planter families, and a number of Asian merchants had also established business operations in Mbozi.

Return of some of the ex-German farms to the Native Authority, and consolidation of others into economic units brought the total number of estates to 40 in the middle 1950s. Since then, the European population has declined with a large outflow at independence, while that of the Asian has probably increased as they expand from merchant activity to farming (Table 2.3). The history of the European and Asian in Mbozi is explored in detail in Chapter 6.

A combination of population pressure in Rungwe District (east of Mbozi) and job opportunities on the estates attracted enterprising Nyakyusa and Ndali men to Mbozi. In 1938, some 1800 were regularly employed on the Mbozi coffee estates, 65% of which were Nyakyusa (Tanganyika ARPC 1938:82). At that time, the immigrants asked to have their own chief established in Mbozi, a request that was officially ignored. The flow of immigrants became a flood, and in 1938–1940 about 10% of the taxpayers in Unyiha were Nyakyusa or Ndali (Hall 1945).

The large numbers of "alien natives" present in various tribal areas in Mbeya District and the local unrest caused by their presence led, in 1944, to a Native Authority Ordinance authorizing the prohibition or restriction of settlement of alien natives by the Native Authorities (Tanzania Archives File Acc 157 L

Table 2.3

Alien Population–Mbeya District

	Population[a, b]	
Nationality	1952	1957
European[c]	596	909
Indian and Pakistani	1019	1253
Goan	39	59
Arab	2	12
Somali	44	79
Colored	38	57
Others	12	8

[a]Data are from Moffett 1958; Tanganyika Non-Native Census, 1958.

[b]Separate data for Mbozi are not available.

[c]There were about 200 Europeans in Mbozi in 1957, including those on estates, in government services, and on missions.

2/11). This provision applied to Umalila, Unyiha, Usongwe, and Usafwa, and was obviously directed toward Nyakyusa emigration. Tanzanian Archives Files (33//23/10 A2/2) contain a number of letters from settlers on the issue of alien Africans. Some of the letters contain copious praise for the Nyakyusa as being skilled, industrious, sober, reliable, and honest. Other letters ask for government aid in removing undesirables among the aliens who were squatters on the estates but refused to work.

Apparently, measures to counteract flow of immigration were unsuccessful, since by 1957 20% of the population of Unyiha consisted of Nyakyusa and Ndali (Table 2.4). The chiefs and headmen apparently were unable to realize the legal power they or potential neighbors had of vetoing a prospective settler (Brock 1966:29). The movement continued in the last decade, but its magnitude is uncertain. Since land has to be obtained through

Table 2.4

The Mbozi Plural Society: African Population, 1957

	Number of people in each Chiefdom[a]		
Origin	Unyiha	Unamwanga	Uwanda
Nyiha	47,911	234	660
Nyika	934	476	211
Lambya	815	205	
Namwanga	1,270	19,298	1,225
Wanda			6,989
Fipa			630
Wungu			471
Mambwe			153
Nyakyusa	4,698		103
Ndali	8,295		
Kinga	396		
Safwa	384		
Malila	323		
Bemba	230		
Nyasa	206		
Others	823	394	314
Total African Population	66,285	20,607	10,756

[a]Data are from East African Statistical Department, 1958.

village political leaders and unaccounted for land is increasingly scarce (Chapter 5), it is likely that this flow is tapering to a halt.

The plural society in Mbozi has had a profound effect on the process of change. In addition to the alien creation of an entire realm of economic opportunities and pressure for development facilities that might otherwise have been realized only much later (Chapter 6), both alien and African elements of the plural society exposed the Nyiha to new ideas, techniques, and aspirations with great intensity. These contacts, as we shall see, have proven to be a major impetus to agricultural change.

Migration and External Interaction

According to a sample survey completed in 1954 (Gulliver 1955, as quoted by Brock 1966:26), more than half of the Nyiha men had never left Mbozi to seek employment. However, at that time one adult man in four was absent. Many of them were young, unmarried men in search of cash and goods, usually to meet the bride-price. In the farming survey I completed in Wasa, Nambinzo, Wanishe, Iyula, and Nyimbili VDCs in 1967, some 77% of all men had traveled and worked outside Mbozi (Table 2.5). Data from modernization surveys conducted in 1967 (Appendix 2) suggest that the present rate of male migration is only about 10%, both in Unyiha and all of Mbozi (Table 2.6). A number of factors explain the recent decline in labor migration. Access to the important Johannesburg labor market has been closed since 1961; the Lupa goldfields which once attracted numerous Nyiha have been out of production for two decades; sisal estates have been in a state of depression for several years; and migration to Zambia is at present officially discouraged in favor of local development. Moreover, local opportunities in cash crop production, employment on estates, and establishment of small businesses have made migration less attractive in recent years. Yet in 1967 an estimated 2617 Mbozi men were away at work, including 1646 Nyiha.

The three major destinations of Mbozi migrants have been the coastal sisal estates, Johannesburg mines, and the Lupa mining area. Smaller numbers were attracted to other opportunities listed in Table 2.5. Information about these opportunities seems

Table 2.5

Nyiha Labor Migration

Destination	Number of men making one or more successful trips[a]	Percentage[b]
Coastal and inland sisal estates	22	28
Johannesburg mines	11	14
Chunya (Lupa mines)	10	13
Zambia: Copper Belt mines and urban areas	5	6
Northeast Tanzania tea estates	5	6
Lake Rukwa fisheries	4	5
Moshi–Arusha sugar estates	4	5
Zambia: European farms	4	5
Usangu rice farms	3	4
Urban employment	3	4
Zanzibar–Pemba clove farms	2	3
Railways and trucking	2	3
Rhodesian farms	2	3
Tukuyu tea estates	1	1
King's African Rifles (World War II)	1	1
Total number of men in the sample	73	
Number who have traveled outside Mbozi for employment	56	
Percentage who have traveled	77%	

[a]Data from a survey completed by the author in 1967 includes only men presently in Mbozi, not those away at that time.
[b]Percentages total to 101 due to rounding.

to have been disseminated in several ways. Many opportunities were made known by word of mouth. Government labor recruiters actively sought employees for the sisal estates, and several Mbozi men claim to have been conscripted into service on them during World War II. Recruitment was most active for the Johannesburg market. Recruits were flown from Mbeya to the Rand and guaranteed a job and passage home. Nyiha who have worked in the Rand view the closed opportunity with mixed feelings. The long, 6-day weeks underground were not pleasant,

Table 2.6

Mbozi Area Labor Migration, 1967[a]

Division	Average number of labor migrants per *Kumi-kumi*	Estimated total number of migrants
Iyula	1.34	398
Igamba	1.17	723
Vwawa	0.83	525
Msangano	2.04	471
Ndalambo	1.56	307
Kamsamba	0.83	247
Mbozi Area	1.18	2671

[a]Data are from the Mbozi Economic Survey, Appendix 2.

yet laborers were able to accumulate significant funds since room and board were supplied. Many Nyiha remember the "color bar" and its ramifications, and some understand why Johannesburg jobs are no longer available. A large proportion of Nyiha men report one or more abortive trips to find work in addition to those in which they were successful. Often a job found was different from that originally sought.

Job migration has meant the bringing of money, ideas, and accumulated goods back to the Mbozi home. The typical Nyiha man is fairly well traveled and has "been around." He has probably seen more developed areas of Tanzania. If not exposed to cash-crop production on an Mbozi estate, he has participated in this sector elsewhere. Having lived outside his home area, he has grown accustomed to at least short-term dependence on market goods. This dependence persists as sugar, salt, edible oil, clothing, soap, razor blades, and the like continue to be purchased rather than locally manufactured or bartered. Labor migration, then, contributes to the overall process of change by increasing exposure to new ideas, technologies, and aspirations.

Settlement Patterns

In response to the widespread unrest during the 19th century, Nyiha constructed fortified, palisaded villages incorporating both people and livestock (Brock 1966:4). Descriptions of the villages provided by Thomson (1881), Kerr-Cross (1890), Johnston (1890), and Fülleborn (1906) suggest they were crowded and unsanitary. Houses in the villages were of two types. The beehive-style *intuma* (Figure 2.1) was common, as was the wattle-and-daub *inyumba* (Figure 2.13).

After German pacification, the Nyiha moved into dispersed homesteads adjacent to areas under cultivation, analogous to widespread abandonment of hill settlements in other parts of Africa (Gleave 1966). Ready access to fields and avoidance of the consequences of crowding may have been important factors for dispersion. The *intuma* housetype has now almost completely disappeared. Aerial photographs made in 1949–1950 indicate a majority of houses were still the circular *inyumba*, but by 1967, even this style was rare. It has been replaced by the more spacious *ibande* (Figure 2.14), which too is a transitional type, as wealthy Nyiha are now constructing European-style houses of brick and corrugated iron (Figure 2.15).

Figure 2.13. The *inyumba* or *inyumbanyumba* (connoting by repetition the round rather than rectangular wall structure), the traditional Nyiha house. Of wattle-and-daub construction, the house has a thatched, overhanging roof. Also visible are grinding stones *(uluwala)*, gourd and clay containers, and traditional stools. The door lock is a blessing of modern time.

Figure 2.14. The *inyumba* or *ibande* (after the Kiswahili *banda*). This rectangular, hip-roof house type has become ubiquitous in Mbozi since 1950. It is derived from both a European-settler model and coastal house types seen by labor migrants. A pile of beans *(imponzo)* dries, awaiting threshing.

Figure 2.15. A European estate house in Mbozi. This type now serves as a model for Nyiha construction of sun-dried mud-brick or burned-brick houses with corrugated metal roofs. A rain gauge adorns every estate lawn.

Modern Nyiha farmsteads *(inkaye)* consist of several dwellings *(ibande)*, grainstores *(ishanga)*, an open courtyard, housing for livestock, and accoutrements of coffee production and food preparation (Figures 2.16 and 2.17). Fields of bananas, maize, and beans, and sometimes a vegetable garden, all in continuous cultivation, are close by. Most staple food is produced in fields farther from the homestead in the traditional agricultural systems. Often, the garden is some distance from the house, along a stream where irrigation is possible. The visitor will be struck by the cleanliness and order in the homestead, and will be greeted by reticently inquisitive women and children, barking but cowardly dogs, curious chickens, and the ubiquitous cat.

Agricultural produce is stored either in the house (beans and

Figure 2.16. A modern Nyiha farmstead *(inkaye)*. There are four houses *(ibande)* here; the fourth is hidden in the far right. The site is in a constant state of evolution; one house is decaying and will be replaced (the average life of a house is 6 to 7 years). An outdoor kitchen was added to the eaves of the central house. A standard allotment is a house for each wife; sometimes a separate house for the husband; and separate houses for unmarried boys *(ivanza)* and girls *(ilele)*. The courtyard around the house *(uluwungu;* compare with *uluwungano,* meeting) is swept clean and has the tanks *(ishivundishililo)* for fermenting coffee. Note the coffee-planting immediately below the farmstead. Often the rows of coffee nearest the buildings are conspicuously larger and healthier than the rest of the planting because of crop and household wastes thrown there.

Figure 2.17. A Nyiha farmstead near Ng'amba Mountain. This African smallholding is between Ngamba and the coffee estate shown in Figure 1.6.

other pulses, ripening bananas, fruits) or in special structures (maize, finger millet, sorghum). The shape of grainstores *(ishanga)* constructed in Mbozi is typical of many peoples in southwestern Tanzania (Harris 1941). The store is cylindrical, set on stilts, with a thatched, removable cover. It is commonly plastered with mud inside and out. Often, the store is woven from sorghum or maize stalks, branches, reeds, or bamboo (Figure 2.6).

Land Tenure

One of the most important aspects of society's relation to environment is the way in which resources are allocated among producing units requiring them. Thus, the allocation of land resources for agricultural purposes is a major facet of organization which people impose on the space around them. Traditionally, among the Nyiha, under little or no pressure on land resources, access to land depended only upon one's remaining on good terms with his neighbors. Brock (1969) thus suggests that in the precolonial period no formal allocation of land was undertaken. Rather, land was available to all, so long as one did not impose on rights already established by others in using a piece of land. This basic right to access based upon use of land not previously cultivated persisted through 1967. At that time, Nyiha informants asserted the right of everyone to have land. Without specific sanction of political officials, one could clear land with the critical proviso that no others held rights over it. Once the land had been cleared, a continuous right is held by the clearer, even while the land is idle or fallow. Anyone wishing to use such land must obtain permission of the last person having cultivated it, if his identity can be ascertained. In addition, a large amount of temporary borrowing and lending of land was common at that time.

During the colonial period, Brock suggests that allocation of land became more formalized, requiring the agreement of the village headman acting on behalf of the chief (Brock 1969). Since land was not particularly scarce, the major consideration seems to have been whether the applicant could gain proper identification with his prospective neighbors, either through descent or as a potentially good neighbor. My supposition is that this more formalized practice developed during the arrival of immigrant peoples and during the period of expansion of Nyiha from the core area toward some of the more peripheral chiefdoms; that is, to areas beyond the locale of their immediate family. This formal allocation of land was less clearly recognized in 1967, due both to the demise of the local chief and headman system and to the evolution of increasing tenacity of land tenure in most of Unyiha.

Neither the traditional right to land through use nor the evolved practice of allocation included any provision for the purchase or sale of land or for permanent transfer without the chief's approval other than through inheritance. Security of tenure was not problematical—once rights to land were obtained, these rights could not be revoked so long as the land was wanted and used. Thus security of tenure was sufficiently great that perennial crops such as coffee were planted without anxiety. By 1961, Brock (1969:5–6) observed that improvements upon and attached to land were widely sold, including coffee plantations and buildings, although land itself could not be bought or sold. In 1967 this remained common in a normative sense, but transfers were beginning to occur in which the distinction between sale of improvements and sale of land itself was obscured. In addition, continued growth of population, allocation of land to cash crops, and the exacerbating effect of land removed from potential Nyiha use for coffee estates and forest reserves all contributed to a virtual enclosure of land in Unyiha. This process and some of its implications are explored more fully in Chapters 5 and 8.

In addition to the allocation of land resources among men as the representatives of households, land resources are also distributed within households. Among the Nyiha in 1967 it was typical to find that the woman (or women in polygynous families) was allocated a portion of the land for production of foodstuffs for herself, her children, and husband (or a portion of his needs in the polygynous case). In addition, a man himself maintained control over and provided the labor for a separate set of fields, including most cash crops. In some cases his food production would be sold; often it would be pooled with the household's food resources and be stored in a common grainstore. Sale of produce from the woman's field required joint consultation with the husband, while he was relatively free to sell crops that exceeded his family's needs. Considerable variation occurred in the pattern of sharing work and produce among polygynous wives, as well as between husband and wives. Extremes included four very independent co-wives from whom the husband *purchased* eggs to present as a gift, to two uninherited co-widows who had worked cooperatively at maintaining their fields and families since the death of their husband. Typically, cordial relations among co-wives required pooling of labor on the fields as well as food at the hearth. Sharing work also provides companionship during such onerous burdens as weeding. Nevertheless, each Nyiha farm will have identifiable land belonging to husband, wife (or wives), or in fallow.

Before examining further the character of Nyiha farming activities, let us turn to the environmental milieu within which agriculture is practiced, to the ecology of agricultural production, and finally to recent agricultural evolution.

Ethnogeography and Geography

chapter **3**

"*Intemo jane mpangula masitu: ahasana.*"–
With my ax I destroy the forest: heat.
NYIHA RIDDLE (Busse 1960:139)

Human societies organize the use of their world's resources not
according to total knowledge of ecological reality but through
knowledge limited by three factors: cultural filters representing
the set of shared terminologies, categories, knowledge, and
beliefs passed from generation to generation by the process of
enculturation; concurrent learning processes or knowledge
becoming, but not yet incorporated as, a part of culture; and
individual and collective levels of insight and aspiration. The
agricultural systems of the Nyiha and their neighbors thus reflect
cultural knowledge accumulated through time, knowledge mod-
ified through contact with other cultures, and progressive learning about the environment they occupy.

The whole set of knowledge and beliefs about the empirical
world held by a cultural group can be termed their ethnoscience

or folk science.[1] If we succeed in isolating one subset or domain of that knowledge, we can examine, for example, the ethnogeography or folk geography of a people, attempting to create verbal or graphic models of their beliefs which are consonant with their understandings. Because this shared body of empirical knowledge and belief helps us to understand much human resource-using activity (and thus the landscape, its residual artifact), we cannot avoid particular attention to this system of cognitive knowledge in discussing a people's agricultural system.

Similarly, we cannot and should not avoid the understandings of our culture, articulated as science. Without at all asserting that our system of science is the sole path to truth, we can map or interpret a folk science within our own understanding, elucidating or amplifying one or even both systems of knowledge. We will argue later that this cross-cultural mapping process may be critical for introducing change—both in the facilitation of change and in minimizing the risk of introducing fallacious knowledge—but here our task is to initiate the dissection of Nyiha environmental knowledge from both points of view, from the Nyiha perspective, and from that of the Western, scientific observer. In the ideal, Nyiha cognition concerning all aspects of their agricultural system would be presented. At the time of my field session, however, I had only limited experience in cognitive research which, combined with a reliance on an interpreter, made such a wide-ranging approach impossible, especially in the face of many other observations also being undertaken. Thus my queries were largely informal and linked to environmental questions—attempting to place agricultural practice within its ecological setting. Nyiha beliefs, such as the use of environmental objects as indicators of agricultural potentials, are presented here within terminological and preliminary theoretical frameworks. Categories of beliefs discussed are suggestive rather than fully reflective of Nyiha thought. Interwoven with interpretations of our own science, this discussion provides one critical perspective in our analysis of change (Knight 1971a).[2]

---

[1]For further discussion of folk science see Blaut 1970; Colby 1963, 1966; Conklin 1957, 1967; Frake 1964; Knight 1971a, 1972; Sturtevant 1964; and Tyler 1969. An excellent case study has been completed by Gladwin (1970).

[2]Formal elicitation and modeling of the Nyiha scientific system provides an important direction for future research. In addition to environmental cognition, Nyiha knowledge in related social and economic areas only briefly explored here should be pursued.

Nyiha understandings of the environment they occupy can be approached through a number of terminological domains. Among the most significant of these is the traditional calendar reflecting cyclical changes in climate and related landscape features through the year, alluded to earlier (Table 1.1). The calendar reflects the seasonal ebb and flow of agricultural activity in response to seasonality in the environment. The Nyiha also have a classification of landscape morphology, of soils, and of vegetation. They are aware of the spatial distribution of macroclimatic resources and the microclimatic modification of them in terms of moisture availability for crops. Finally, Nyiha know the sources of environmental risk, exceptions to their basic expectations with which they must cope at uncertain frequencies.

The information discussed in this chapter is derived from three sources. First, the Nyiha themselves expressed their ideas about environment and agriculture in innumerable formal and informal interviews conducted with my interpreter in 1966–1967. Second, comparative "scientific" data were derived from a concurrent geographical reconnaissance of Mbozi. Finally, selected terminology in Shinyiha is presented drawing upon Busse's work on the Nyiha language (Busse 1960).

Climate and the Traditional Calendar

Nyiha agricultural activities are closely tied to the marked seasonal rainfall regime in Mbozi. Nyiha distinguish two major seasons each year (Table 3.1). *Ishisiku*, the rainy season, lasts from October or November through April or May. The dry season, *ishisanya*, includes the months from May through October. The definition of these seasons *vis-à-vis* our calendar is flexible. *Ishisiku* begins when the rains arrive, sometimes in October, but usually in November. *Ishisanya* begins when the rains end. Within the two major seasons, *ulupepo* (the cold season) and *mundundu* (the early rains) are distinguished. *Ulupepo* is the low sun period ("winter") when the wind shifts from the southeasterlies which bring rain to the northeasterlies which are dry. The warmest time of the year is October to November, when the sun is rising higher in the sky and rain clouds have yet to abate the receipt of solar radiation.

Ishisanya is the period of preparation of new fields, especially cutting of woodland for *nkomanjila* (Chapter 4). In *mundundu*

Table 3.1

Nyiha Calendar[a]

Seasons	Months[b]	Related terminology[c]
Ishisiku	*Umwene*	*Mwene:* chief; *kwenye:* to look ahead
	Umukuwe	*Ikuwe:* a plant flowering in February, indicating time for planting finger millet is past
	Umufingwe	*Mufingwe:* something unopened; tasseling and flowering of maize
	Umuhanda	*Kuhanda:* to splash or squirt; to mix water and soil to make houses; mixture of activities
Ulepepo	*Umupangwe*	*Impangwe:* a plant (*Pentas decora* var. *triangularis*) flowering in May as are most crops
	Uholo	*Aholo:* moist, undried; crops ready to be eaten green
	Uhavunasote (*Upenza*)	*Kuvuna:* to harvest; *usote:* a creeping plant (*Mukua maderaspatana*) eaten as a vegetable; *kupenza:* to move or take off something to see inside, opening of finger millet
Ishisanya	*Upukutu* (*Uhatonyia*)	*Kupukuta:* to lose leaves or let something fall down; *kutona:* to ripen or bear fruit
	Ulavila	*Kulavila:* to rise early, refers to grass shoots and tree leaves
	Usanyikale	*Ulusanya:* heat; *kali:* hot or sharp; *ahasana:* heat
Mundundu	*Umwanga*	*Umwanga:* a bird whose arrival announces the coming of the rains; *umwanja:* light, both sunshine and rain, indicates time of planting
	Ulima	*Kulima:* to cultivate; *amalimo:* hoeing or field work

[a]The calendar was recorded from informants in the field in 1966–1967.
[b]Months listed are from January to December.
[c]Related terminology is from informants and from Busse (1960).

the activity of field preparation reaches a peak. Hoeing, weeding, and making of bean, cassava, and other root-crop fields takes place during *ishisiku*. In *ulupepo* the main crops come to maturity and are harvested.

Within the major seasons, Nyiha recognize 12 months, each having traditional names which can be etymologically analyzed (Table 3.1). Whether the division of the year into 12 months is wholly traditional is uncertain, although the Nyiha claim it to be. However, the terms used are clearly related to ecological events happening during the month and to corresponding human activities. For example, *Ulima* (December) is the month of hoeing of major crops, and the verb root, *–lima,* means *to cultivate.*

The rainfall of Unyiha is both plentiful and relatively reliable. Over much of the central Mbozi Plateau, average seasonal totals of 50 inches or more are received (Table 3.2). The lowest seasonal total recorded in the last 25 years was 39 inches in 1948–1949 at Mlowo Farm (Estate 30 on the front end papers).

Table 3.2

Mbozi Climatic Data

| | Values at Mbimba Experiment Station[a] | | | Mkulwe Mission[b] |
| | Rain | Temperature | | |
	Mean rainfall, 1956–1967	Mean minimum °F	Mean maximum °F	Mean rainfall, 1945–1964
January	9.25	58.5	75.5	7.57
February	8.38	58.9	75.8	6.06
March	10.06	58.4	75.7	5.80
April	7.25	57.3	76.0	2.77
May	1.33	52.7	76.7	0.54
June	0.15	49.3	76.3	0.00
July	0.02	48.0	76.3	0.00
August	0.02	48.3	78.3	0.03
September	0.30	48.9	81.9	0.05
October	1.43	51.5	83.4	0.27
November	3.99	55.2	80.5	2.43
December	8.95	57.6	76.3	6.72
Year	51.13	53.7	77.7	32.24

[a]Values shown were calculated from data on file at the Mbimba Agricultural Experiment Station.

[b]Data for Mkulwe Mission in the Rukwa Valley are from the East African Meteorological Department (1965).

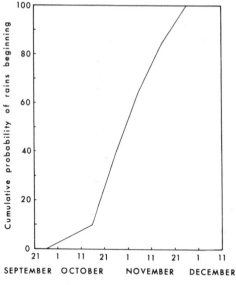

Figure 3.1. Initiation of the rains. The cumulative probability of the rains having begun before or during a given 10–day interval, using 0.67 inches of rain within 10 days as indicating the rains have begun.

Since Mlowo Farm and the whole of the eastern hills usually receive an average of 5 inches more rain than the plateau proper, totals of about 34–35 inches were probably common. Since 1950, the lowest seasonal total at Mbimba was 36.68 inches in 1953–1954. The highest seasonal total recorded at Mlowo Farm was 77.59 inches in 1961–1962; at Mbimba 69.04 inches were received in the same season. Using an arbitrary rainfall of 2 inches per month or 0.67 inches per 10-day-interval as a criteria for initiation of the rains, the rains most frequently begin in late October or early November (Figure 3.1). Often, the initial rains are followed by a break of a week or two with little or no rain before the continuous rains commence. The rainy season virtually always ends in the last 2 weeks of April or first 2 weeks in May, with a few scattered showers later.

While rainfall is usually plentiful and reliable in Unyiha, in the drier areas to the northwest rainfall averages are lower and relative variability greater. In the dry season of 1949, Lake Rukwa dried up almost completely (Moffett 1958:266), reflecting low rainfall receipts both in the valley itself and in its drainage basin. The average rainfall at Mkulwe Mission is an annual 32 inches, and during the seasons 1957–1958 to 1964–1965, when the average seasonal rainfall was 37 inches, a low of 25 inches was recorded in 1964–1965. In both that season and in the 1966–1967 season, lack of expected midseason rainfall caused significant crop failure.

Nyiha recognize the increasing aridity toward the northwest. The significance of this gradation is most obvious in coffee, and many of my informants in Nambinzo VDC held the unfulfilled hope that I could solve the dilemma of rainfall they clearly saw as too scanty for coffee production. With the exception of one interview in Wanishe, only informants in Nambinzo listed drought as a most important agricultural risk. Hence, Nyiha perception of climatic resources has contributed to a slackening of migration to the plateau margins as Nambinzo and the northwest in general has proved marginal for coffee, the major cash crop. Nyiha knowledge of water resources, however, is not limited to rainfall receipts, as will be seen presently in a discussion of soils.

Traditional Land Types

Traditional Nyiha land terminology fits into a general taxonomic relationship (Figure 3.2). At the broadest scale, *imboto* (fertile or cultivable, inhabited land) is distinguished from *mwilala* (bush, uninhabited wilderness). The central Mbozi Plateau is *imboto* while northern sections of Itandula, Magamba, and Isansa VDCs are *mwilala*. The Nyiha have a term *mulukwa* referring to hotter, drier lowland areas in contrast to Unyiha, for which there is no contrasting generic term.[3] Hills or mountains including inselbergs are *magamba*, while flat areas or valleys are *ulutalama*. Cultivable hills, *magamba*, are contrasted to hills with bare rock and stones, *ivinkarango*. *Kunjenje* refers to stream-dissected areas as opposed to *kumahombe*, the gradually sloping areas of the plateau surface. *Muntalama* (valley) is contrasted with *munzi* (upland), which can also be set in opposition to *ihombe* (mbuga grassland). Within *kunjenje*, *ishihoto* refers to a small, steep, stream-head valley; the slopes in *kunjenje* are termed *ulwiha* or *uluvambala*. Hydrologically speaking, *aminzi* is water; *ilyinso*, spring; *ividwima* or *ivintawala*, flooded areas or *ihombe* while flooded; *itinkinya*, swamp; and *injenje*, river.

[3]It is interesting to compare *mulukwa* with Lake *Rukwa*. R and L are often confused in transliteration; the Nyiha term may simply be a generalization of the specific area name.

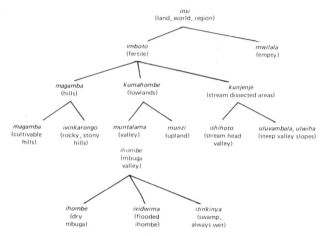

Figure 3.2. Nyiha land classification.

All of these terms have agricultural connotations. The most important distinctions in relation to soils are mentioned below. For cropping in general, the *munzi–ihombe* dichotomy is most frequently mentioned, second only in importance to the vegetation dichotomy between *ishihombe* or *ihombe* (grassland) and *indunguti*, *invuti*, or *intundu* (woodland, regrowth woodland, and bush). Fallow land of all descriptions is known as *ukutuzye*.

Soils

The soils of central Unyiha have been studied in detail by Spurr (1955). More general studies of East African soils (Scott 1962) and Tanganyika soils (Anderson 1963) provide smaller-scale studies of Mbozi Area within larger contexts. Figure 3.3 is a map of the soil–topography associations of Mbozi Area following Scott's (1962) classification scheme with boundaries revised from Scott's map in the field. A detailed soil map of central Unyiha was provided by Spurr (1955). Our purpose here is to describe briefly the general features of soils occurring in Unyiha and indicate the Nyiha understanding of soils in relation to these existing studies.

With the exception of rocky inselbergs which cannot be shown at the scale of Figure 3.3, the soils of Unyiha fall into three broad classes. The well-drained upland soils are dark brown loams derived from volcanic ash overlying older soils which are friable red clays. These clays constitute the surface soil on

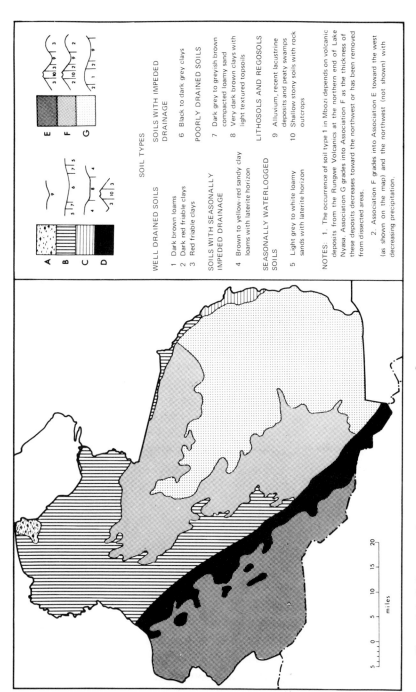

Figure 3.3. Soil–topography associations of Mbozi Area. [The classification scheme is derived from R. M. Scott, *Soils*, a map accompanying E. W. Russell (1962).]

SOIL TYPES

WELL DRAINED SOILS

1 Dark brown loams
2 Dark red friable clays
3 Red friable clays

SOILS WITH SEASONALLY
IMPEDED DRAINAGE

4 Brown to yellow-red sandy clay
   loams with laterite horizon

SEASONALLY WATERLOGGED
SOILS

5 Light grey to white loamy
   sands with laterite horizon

SOILS WITH IMPEDED
DRAINAGE

6 Black to dark grey clays

POORLY DRAINED SOILS

7 Dark grey to greyish brown
   compacted loamy sand
8 Very dark brown clays with
   light textured topsoils

LITHOSOLS AND REGOSOLS

9 Alluvium, recent lacustrine
   deposits and peaty swamps
10 Shallow stony soils with rock
   outcrops

NOTES: 1. The occurrence of soil type 1 in Mbozi depends on volcanic
deposits from the Rungwe Volcanics at the northern end of Lake
Nyasa. Association G grades into Association F as the thickness of
these deposits decreases toward the northwest or has been removed
from dissected areas.

2. Association F grades into Association E toward the west
(as shown on the map) and the northwest (not shown) with
decreasing precipitation.

5 0 5 10 15 20
miles

A
B
C
D

E
F
G

slopes where the overlying volcanic soils have been removed. Finally, there are broad areas of alluvial-meadow and valley-bottom soils which are often under water during the rains.

The volcanic material which strongly influences central Unyiha soils gradually disappears toward the north and west. This airborne volcanic ash was derived from Pleistocene to Recent volcanic activity at the northern end of Lake Nyasa (Harkin 1960). Shallow, stony soils characterize steep slopes and sharp interfluves on the plateau margin. The dark brown loams become less frequent and red clays predominate in upland areas.

The Nyiha classification of soils can be directly compared to the general soil types mentioned. The major soil types are *inkanka,* the red friable clays; *inyilu,* the upland loams of volcanic derivation; *insusu,* locally distributed ash-colored silt loams that represent a less well-developed volcanic-ash soil; and several minor soil types, *inzeru, lusenga, iwumba,* and *ibanya.*

The red *inkanka* soils are typical of most steeply sloping areas *(uluvambala)* of central and western Unyiha, and indeed, most of the dissected plateau margin *(kunjenje).* These are well-drained, fertile clays and clay loams in which a laterite zone of iron concretions is found only at depths over 6 feet, if at all. *Inkanka* soil has a moderate moisture-holding capacity and is considered by the Nyiha to be suitable for most crops. It is indicated by the major *miombo* woodland species, *ilenje (Brachystegia* sp.), *ing'anzo (B. boehmii),* and *insani (B. bussei).*

The *inyilu* soil type is less ubiquitous than *inkanka* except in the east where it is dominant. *Inyilu* is considered to be the best soil for most crops. The soil term is derived from *uvwilu,* black. It is found on most gently sloping upland areas and is indicated specifically by *iwula* trees *(Parinari curatellifolia).* It is usually underlain by *insusu* or *inkanka. Inyilu* is especially valuable because it is both fertile and holds a large amount of moisture. Rainfall soak-in on it is rapid. It can be used for virtually any crop and is especially suitable for coffee.

Insusu soils are lighter in color than *inyilu* and are the last soils to dry in the dry season. They are found only in eastern Unyiha and usually on gentle to medium slopes *(ulutalama),* in valleys, and on hillsides. *Insusu* is considered to be less fertile than *inyilu,* but is the best soil for cultivation of sweet potatoes because it stays wet well into the dry season. *Insusu* is very dusty when dry, but rapidly absorbs rainfall. It is indicated by *ikusu* trees *(Uapaca kirkiana)* and under grassland is usually cultivated using the *nkule* method (Chapter 4). The *inyilu–insusu* soils at Mbimba have an available water capacity of an amazing 7 inches in the first 2 feet and an additional 8 or 9 inches in the next 2 feet of depth (Mbimba Station Files).

Inzeru soils are found in *ihombe* areas although the term is applied to droughty *inyilu* (darkly colored) soils elsewhere. *Inzeru* soils are indicated in upland by *ilaji* trees *(Brachystegia spiciformis)* and the woody herb *ihokwa (Commelina* sp.). They are not considered as very good for crops because they dry out too rapidly. Usually underlain by *iwumba,* the clay used in making pots, *inzeru* soils in *mbuga* areas have a lower horizon of *ibanya,* the white soil used in "painting" houses.

Finally, *lusenga* soils are sands found at the base of inselbergs *(magamba)* and along streams *(ishihoto).* In western Unyiha, the *inkanka* soils have a markedly lower moisture-holding capacity than in the Vwawa area. Here, Nyiha do not hesitate to classify a soil descriptively as a combination of basic types. The valleyside soils of the plateau margin, then, are often called combinations of *inkanka* and *lusenga.* People in western Wasa, for example, claim they receive no less rainfall than to the east, but less is held by the soil, a condition we would term "droughty." Thus, a major dimension of Nyiha knowledge of soils concerns the soil climate.

The soil constitutes a kind of repository in which moisture is deposited by precipitation and withdrawn by gravity, plant transpiration, and evaporation from the soil surface. We can understand much of the operation of available soil water resources by reference to the combined effects of evaporation and transpiration, known as evapotranspiration (Thornthwaite 1948; Chang 1965). Evapotranspiration represents total water loss from a soil/vegetation ecosystem due primarily to atmospheric parameters affecting evaporation—energy, humidity, and wind. Using these meteorological parameters to estimate potential evapotranspiration, a very simple budgeting procedure then allows us to model the status of soil moisture resources through the year. For each time period, we add to the soil precipitation received and withdraw needed evapotranspiration, expressing the resulting balance as either an addition to soil moisture storage, runoff (when the soil's capacity for moisture is filled), drawdown of the soil storage, or actual water needs not supplied (deficit). The major effect of the soil itself in this budget is that its texture determines to a large extent the amount of moisture it can hold within the root zone of typical crops.

The average annual water balance calculated monthly at Mbimba and for contrast, Mkulwe Mission, can be presented diagrammatically (Figure 3.4). The calculations were made from climatological data gathered in Mbozi, with the Penman method as described by McCulloch (1965) used to calculate potential evapotranspiration. Several simplifying assumptions were made in creating these examples. First, it was assumed that the soil moisture storage capacity (available water) in the root zone of a hypothetical crop was the equivalent of 8 inches of rainfall at Mbimba and 4 inches in the loamy sand at Mkulwe. Second, it was assumed that soil moisture was withdrawn at the potential rate until that stored in the soil was exhausted. Finally, there was no alteration of the potential evapotranspiration by season or crop stage. With these simplifications, the diagrams illustrate the seasonality of soil moisture resources. On the average, precipitation more than supplies potential soil water losses at Mbimba from November through April. Once the soil is fully charged, additional excess moisture becomes direct runoff or enters the ground water table. In this simplified model, precipitation at Mkulwe is suboptimal in all months except January.

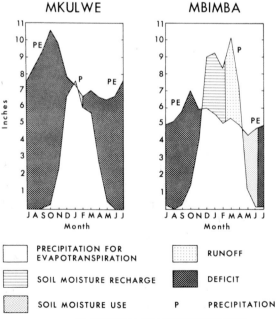

Figure 3.4. Water balance for Mkulwe and Mbimba.

The effect of a droughty soil on the soil moisture balance will be to shorten the soil moisture growing season since there will be a smaller amount of moisture to be tapped when potential evapotranspiration exceeds precipitation. Hence, the Nyiha contention that western Wasa and Nambinzo are drier not only because of less rain but also because of droughty soils is scientifically valid. In the troughs and Rukwa Plain, higher temperatures mean higher potential evapotranspiration. This, plus the lower moisture-holding capacity of the soils exacerbates the effect of low rainfall.

Dagg (1968:3) has observed that runoff may be deceptively high in the tropics. In central Tanzania, runoff from a sandy loam with a minimal 1% slope can amount to 40% at moderate rainfall intensities. Such a proportion of runoff may be critically important in areas of marginal rainfall. The kind of soil, its preparation, and its protection against rain impact will affect the amount of soak-in versus runoff. The water balance diagrams presented here are based on the rather unjustified assumption of 100% soak-in. More realistically, the rainfall received should be decreased to account for the direct runoff.

There was some alarm in Mbozi in the early 1950s when after several years of low rainfall, dry-season water shortages became acute. In a report, the Tanganyika Engineering Geologist (Tanganyika Engineering Geologist 1955) suggested that dry season stream failures would become increasingly common unless cultivation and overgrazing in catchment areas were not controlled. By that time, the first forest reserves had been demarcated for just that purpose.

Vegetation[4]

Nyiha are concerned with vegetation on two different scales. On a detailed scale, they recognize, name, and utilize individual or closely related species both as indicators of land potentials for agricultural use and as economically useful plants. For example, couch grass or *usankwe (Digitaria scalarum)* is an indicator of a field only recently abandoned, while the herb

[4]The following additional references on vegetation types and identification may be of interest: Stevenson 1931; Michelmore 1939; Jackson 1940; Eggeling 1947, 1956; Tanner 1951; Burtt 1942, 1953, 1957; Pielou 1952; Dale 1953, 1961; Vesey-Fitzgerald 1955; Backlund 1956; Rattray 1954; Trapnell 1959, 1962; Bogdan 1958, 1961; Gillman 1949; Lind 1962; White 1962; Albrecht 1964; Napper 1965; Dean 1967.

ishisumbwe (Dolichos sp.) is never found on old fields, suggesting that no one has cultivated the area within memory. Many other examples of the role of indicator species are mentioned later in the sections on cropping systems. In 1966–1967 I collected over 175 plants with specific Nyiha names (plus many others), many of which have medicinal, culinary, or economic uses. These plants, with their Shinyiha name and use as well as their scientific nomenclature are listed in Appendix 1.

On a broader scale, the Nyiha distinguish the major physiognomic features of vegetation growing in an area. Their initial choice among alternative cropping systems is based on this classification. The vegetation mapping system used on the Directorate of Overseas Surveys topographic sheets covering Mbozi Area corresponds closely to the Nyiha system. The vegetation map of Mbozi (see back end papers) was compiled from these topographic sheets with further interpretation and revision in the field. The vegetation map is intended to represent the dynamic as well as physiognomic aspects of vegetation in Mbozi. In addition, it reflects large-scale vegetation features as classified by the Nyiha (see also Table 3.3).

In Nyiha terminology, *indunguti* or deciduous woodland is the virgin or fully regrown *Brachystegia* (Kiswahili, *miombo*) wood-

Table 3.3

Dominant and Frequent Genera

Area	Woodland and Scrub	Grassland	Mbuga
Mbozi Plateau	*Brachystegia, Uapaca, Isoberlinia, Strychnos, Pterocarpus*	*Hyparrhenia, Loudetia, Melinis, Andropogon, Cymbopogon, Panicum, Pennisetum*	*Themeda, Hyparrhenia,* Cyperaceae (Sedges)
Mbozi Parinari Woodland	*Parinari, Uapaca, Brachystegia, Isoberlinia, Mangifera* (mango)		
Msangano Trough	*Brachstegia, Acacia*	*Hyparrhenia, Panicum*	
Rukwa Plain	*Acacia, Combretum, Brachystegia, Pterocarpus*	*Hyparrhenia, Chloris, Panicum, Cynodon, Digitaria, Sporobolus*	*Cynodon, Sparobolus, Diplachne, Imperata,* Cyperaceae
Ufipa Plateau	*Brachystegia, Isoberlinia, Combretum*	*Themeda, Hyparrhenia, Exotheca, Loudetia*	*Trichopteryx, Hyparrhenia,* Cyperaceae

land characteristic of much of Tanzania and east-central Africa. Dominant genera in this formation include *Brachystegia, Uapaca, Isoberlinia, Strychnos,* and *Pterocarpus.* In a mature state, this formation has a single contiguous canopy and an open floor. It responds to the moisture stress of the dry season by shedding its leaves and is often inadvertently burned in seasonal firing of grasslands and fields (Trapnell 1959). To the Nyiha, *indunguti* immediately indicates the making of *nkomanjila.* Where there is a particularly dense stand of trees, Nyiha use the term *mwisole.* Much of the primary *miombo* woodland in Mbozi has been cut in the course of human occupation. Areas of uncut *miombo* are those too dry or too rugged for cultivation, including the plateau margin and inselbergs. *Miombo* has succumbed both to agriculture and cutting of firewood and timber.

Intundu or scrub represents an early stage in the succession to *miombo* woodland in Unyiha. Young trees and bushes of *miombo* genera are typical. *Invuti,* a later successional stage, is dominated by trees which have not yet reached their full height. Nyiha will make *nkomanjila* in *intundu* or *invuti* when *indunguti* is not available. Outside of Unyiha, scrub woodland represents more arid conditions, repeated firing, and/or cultivation.

A wooded scrub and grassland mosaic is the more usual circumstance in which scrub occurs. For such a mosaic Nyiha do not have an inclusive term. The grassland portion of the mosaic is known as *ishihombe,* and a particularly tall grassland formation is called *isanga* or *ikanga.* In Unyiha, the scrub/grassland mosaic is a consequence of human activity, representing various stages of regrowth after a field *(umugunda)* has been abandoned to fallow. It is interesting, indeed, that the Nyiha have a general term for grassland, *ilala,* that is derived from the root *-lala,* which in adjectival use means old or tired. In the course of repeated cultivation and firing of the fallow, open grassland is maintained as a fire subclimax. Dominant grassland genera include *Hyparrhenia, Loudetia, Melinis, Andropogon, Cymbopogon, Panicum,* and *Pennisetum. Ishihombe* is a general indicator for *nkule* or other grassland agricultural systems discussed in Chapter 4.

Ihombe or *mbuga* are grassland areas which are flooded during the rainy season. They have three separate vegetation zones. The upper zone has a high water table during the rains and can be used by the Nyiha for fields called *ihombe.* This zone is dominated by *Hyparrhenia* species. *Themeda* is dominant in the middle zone which is flooded during *ishisiku* but dry in *ishisanya.* Areas wet virtually year-round have no particular grass dominant but do include a large proportion of Cyperaceae (sedges). Small perennial swamps *(itinkinya)* occur in the *mbuga* and are often the site for making *vilimbika* (gardens). *Itinkinya* are too small to be shown on a regional map (back end papers); the large swamp in the Rukwa Plain is seasonally flooded by varying levels of Lake Rukwa and the Momba River.

In addition to the types of woodland cited, Nyiha recognize *isitu,* the dense forest that fringes rivers. These forests are too limited in extent to be of agricultural value although bamboo *(ilanzi)* and other useful species may be cut. Because *isitu* are circumscribed, they cannot be shown at the regional scale (back end papers).

In mapping the vegetation of Mbozi, I have included three additional categories. Areas of intense human occupation and cultivation are demarcated. Here vegetation is under active human control. The Mbozi Plateau, for example, could be termed a "cultivation steppe" in which fallow grassland, bush, and cultivated fields form a mosaic, with the proportion of bush decreasing through time. Areas cleared on estates for cultivation of coffee are sufficiently large to be shown at regional scale. Not all of the cleared areas are presently planted to coffee, and there are a myriad of small coffee plots throughout Unyiha. Finally, following the central African Rail Link Survey (Gibb 1952), an area of *Parinari curatellifolia (iwula)* woodland is shown. The *iwula, ikusu (Uapaca kirkiana), impangwe (U. pilosa),* and other trees that are common and sometimes locally dominant in this formation have edible fruits. They are purposely spared in the course of cutting woodland for fields; over time, they with the introduced mango *(inyembe; Mangifera indica)* are virtually the only trees found in densely populated areas.

Environmental Risk and Uncertainty

When asked the open-ended question "What are the risks you face in agriculture?" Nyiha informants replied with the same kinds of phenomena again and again. The most frequent ordering of risks was as follows:

1. Locusts
2. Too little rainfall

3. Too much rainfall
   a. Hail
   b. Rainy season too long
   c. Flooding of *ihombe*
4. Pests including wild animals, birds, and insects other than locusts

In addition to these risks, crop diseases, frost, and lightning occur, but, interestingly, were never mentioned by informants without prompting. It is likely that crop diseases are simply tolerated as a continuous, *expected* occurrence. Only when a rare event such as a rainy season lasting too long results in fungi on grains is disease alarming. Frosts and lightning are sufficiently rare that few people are concerned with them. Lightning (*ulumesyo*) did kill nine cattle in Vwawa VDC in 1965, but this was the only damage report due to lightning I recorded.

The following discussion is divided into biological and climatic risks. In addition to these forms of risk inherent in the Nyiha environment, there are clearly other classes of risk of which the Nyiha are cognizant in agricultural decision making. Both economic and social risks will be discussed within the context of contemporary agricultural practice in Chapter 5.

Biological Uncertainty

One source of biological uncertainty in Nyiha agriculture has been locusts (Allan 1931; Gunn 1956, 1957; Rainey 1957; Bullen 1966). Without exception, they are listed as the major risk to agriculture. Three locust species are of potential danger to Mbozi. The Desert Locust (*Schistocerca gregaria*) is widespread in the Near East and Northern Africa. While it does invade East Africa, it has not been found as far south as Mbozi. The African migratory locust (*Locusta migratoria migratorioides*) is more important. It is distributed widely in Africa. When favorable local conditions prevail, the innocuous grasshopper phase of the locust gives way to the migratory swarm phase which is extremely destructive.

The locally most significant species is the Red Locust (*Nomadacris septemfasciata*). There are two breeding areas of the Red Locust in the Corridor region. One is the Mweru swamp area on the Zambian–Congo boundary. Most important for the

Nyiha is the Lake Rukwa Basin. From relatively limited breeding areas the locust has invaded some three million square miles of Africa since 1930. Gunn (1956:15) noted the general relationship between lake level and Red Locust breeding:

> If the lake level is high, presumably the water table under the grass plains is also high and it then takes little rain to waterlog the ground; such ground is unfavorable to the eggs. If the lake and water table are low, it is then possible for suitable distributed rains to produce a damp soil, comparatively well-drained, to a favorable texture for eggs. Moreover, in these circumstances the area available for breeding is much larger because of the recession of the lake.

Fortunately, prevailing easterly winds have usually carried the locust away from rather than into Mbozi.

Nyiha informants mentioned five locust invasions. Four of them—1924, 1928, 1931, 1940—were attributed to *ushipumi*, the Red Locust, while the 1932 invasion was by *uhalonga*, the African migratory locust. The 1924 incident was the most devastating, and stories of people eating banana stalks and wild plants are still told.

The Red Locust Control Service was founded in 1941 for patrolling of the Rukwa and Mweru breeding areas. Hopefully, increases in locust populations could be detected, and suitable measures taken. However, it was found that swarms could form too quickly to provide control measures in these rather inaccessible areas. In 1949 The International Red Locust Control Service was established with the support of Belgium, Great Britain, South Africa, Southern Rhodesia, and Portugal. The functions of the Service were (Gunn 1957:11):

1. To organize the permanent control of known outbreak areas of the Red Locust and the investigation of regions suspected of being sources of origin of the Red Locust
2. To take steps for the immediate destruction of any incipient swarms discovered in the recognized outbreak areas
3. To organize information about swarms outside the outbreak areas
4. To keep participating Governments informed of the situation, and
5. To study the habits and ecology of the Red Locust and the methods for its control.

Ecological research suggested that control of grassland burning might lead to a concentration of egg laying in burned areas. This was instituted with the result that great savings were realized in locating and controlling the locust. Today, the Red Locust is no longer a menace to agriculture so long as the vigil is maintained. Many Nyiha, however, remember the last locust invasion (or stories about it) and welcome the root crop cassava as an insurance against locust destruction.

Birds such as the red-billed finch, sparrow, yellow weaver bird, bishop bird, and dove cause considerable damage to crops in Africa as a whole. Majisu and Doggett (1968) estimated that in Africa some 10–20% of the annual sorghum crop is lost to birds. Nyiha recognize the importance of bird damage, but realizing there is very little they can do to prevent it, tolerate birds as a nuisance. Small traps *(iliva)* are made to catch birds, and in fields far from homesteads, guard houses are built and continuously occupied as the grains come to maturity. This is a protection against monkeys as well, and is practiced most commonly in the dissected plateau margins in the north and west. Nyiha used to fence their fields but no longer do so. It is unnecessary in central Mbozi, and in the more isolated areas, a common response to the wild animal danger is to have one field that is closely guarded during the growing season.

The plant pest of greatest significance is the maize stalkborer *(Brusseola fusca* Fuller) which also attacks sorghum. Severe infestations of the pest were reported in Unyiha in 1951 (Tanganyika MDR 1951). The parent moth lays the eggs of the stalkborer in leaf sheaths of young plants. Larvae eat the leaf surfaces and subsequently borrow into the stalk by way of the funnel at the base of the leaf. The larvae continue to live in the stalk of harvested maize and sorghum in the field as well as in stalks cut for building purposes. These larvae can pupate and emerge as adult moths at the beginning of the following rains. Insecticidal treatment is not used in Mbozi. Burning of old stalks is not common, and when done, is not with the intention of disease control. However, burying of old stalks in mounds *(matuta),* which is commonly practiced, may also be effectual in preventing emergence of the adult moth. When portable sprayers become more common among coffee growers, the means by which the stalkborer can be effectually and cheaply controlled by insecticides will be available. Spraying might cost 2 or 3 shillings per acre for the chemical, but could result in yield increases of 50–300% or up to 120 shillings per acre (Swaine 1961).

Climatic Uncertainty

Because drought was listed as an important risk by most Nyiha informants, the nature of this risk requires careful examination. Informants were in all cases unable to identify any drought year other than 1949, while locust invasions as far back as 1924 were specifically cited. When questioned more closely, much of their concern about drought involved timing of the arrival of the rains rather than the actual amounts received, suggesting concern with the impact of early rains on seedlings or perhaps coffee production. Moreover, discussion about drought is often focused on the occurrence of this hazard in nearby locales rather than Unyiha. News of famine or actual migration of those stricken could provide justification for Nyiha concern. Available climatic data for Mbozi were examined in an attempt to elucidate Nyiha concern with drought.

Drought risk can be handled in two ways. First, looking simply at seasonal totals expressed probabilistically, the Unyiha rainfall values at any confidence level are adequate (Table 3.4) while those in Unamwanga and Uwanda are clearly marginal. A second and more rigorous approach to the problem of drought was provided by P. W. Porter (1968). Porter has developed a computer program in which monthly water balance status is continuously simulated, appropriate optimal planting dates selected, potential evapotranspiration   corrected according to status of the crop, and the resulting seasons examined as to whether crop yields would have been depressed by moisture deficits. Using water needed by maize, which is more demanding in this regard than most grains, and assuming a soil moisture storage capacity of 8 inches, rainfall data from 1950–1967 for Mbimba were analyzed using Porter's program. The program expresses the significance of drought in any one season by a ratio of hypothetical yields realized to yields expected if moisture were not limiting. For the 16 years for which complete data were available, only one (1951–1952) had less than a perfect score (100) and that was 96. The major way in which farmers might have failed to do as well as the program indicates as if they had planted too early, before the rains sufficiently continuous to maintain adequate soil moisture had arrived. If a farmer had planted in October 1962 his yield would have been only 20% of that of a farmer who

Table 3.4

Mbozi Rainfall Confidence

Percentage of seasons that will have rainfall less than or equal to values shown on right	Rainfall by station[a]			
	Mlowo Farm	Mbimba Expt. Station	Mkulwe Mission[b]	Isoka Zambia[c]
25	47.7	45.3	34.9	37.0
50	54.0	52.4	36.7	42.0
75	62.3	59.0	42.3	46.5
Mean annual precipitation	56.8	51.1	37.0	41.8

[a]Values were calculated by fitting a normal curve to published and unpublished data.

[b]The Mkulwe seasonal average is for 1957–1965; the 19-year average ending in 1964 was 32.2 inches (East African Meteorological Department 1965).

[c]Isoka is near Tunduma and indicates rainfall values of the Ndalambo area of Unamwanga (Rhodesia and Nyasaland, 1957).

waited until November or December and had a perfect season! This is instructive regarding the government plea for early planting mentioned in the preface.

A similar analysis was completed for Mkulwe Mission using both a 3- and 8-inch soil storage capacity. The lower actual water demand by maize during its early and late growth compared to potential evapotranspiration makes Mkulwe appear a little more favorable than does the water balance diagram discussed earlier (Figure 3.4). With an assumed 3-inch storage, 2 of the last 9 years had yield ratios less than 90% according to Porter's model. It is interesting that the predicted famines (1964–1965, 1966–1967) had in fact materialized and governmental relief measures had indeed been required in Uwanda. An assumed 8-inch storage capacity did not significantly alter the model, and the same famines were predicted. In summary, Porter's model mirrors reality very closely in analyzing soil moisture and drought risk. I cannot help but conclude that the Nyiha concern with local drought is not focused upon a real risk to their food production. An alternative hypothesis to account for drought being considered significant would concern the vulnerability of coffee to low rainfall receipts, particularly during the first few months after flowering (October through December in Mbozi). Variation in the early rains results in obvious decreases in yields and cash income. This hypothesis could be partially confirmed in future research if it could be shown that Nyiha men, especially, rate drought an important risk to coffee. Similar queries among women, whose activity focuses upon food crops, should suggest that they view drought as less significant than do men. The former role of the chiefs in time of rain failure, then, would have been focused upon arrival of the rains, rather than prolonged seasonal drought.

While in central Mbozi (Unyiha) drought does not seem an important climatic risk to food crop production, hail *(idjela)* is. Hail can occur any time during the rains, but it is especially significant early in the season. At that time seedlings of the grain crops are subject to damage, and the cash-crop coffee is especially vulnerable during flowering. Informants in Wanishe VDC reported a minor local famine due to hail in 1954; and the 1965 coffee crop, for example, was significantly lower than previous years in areas of Igamba, Vwawa, and Iyula ravaged by hail. Heavy hail will tear banana leaves and strip leaves from cassava but neither event is particularly significant. There were no reports of a mature grain crop having been lodged by hail; under nonmechanized harvesting this would be of little consequence even if it did occur, although even heavy rain *(umulamba)* could be potentially irksome in this respect.

Finally, excessive rains are considered a significant risk when they occur as a result of an extended rainy season or cause flooding of *ihombe* fields at *mbuga* margins. Should *ihombe* fields become flooded, the crop may be damaged or destroyed. Prolonged rains increase the risk of disease in grain crops and prevent adequate drying of grain for storage. It is likely that extended rains will be of greater significance as cultivation of late season wheat becomes increasingly common, although the gains to coffee by more rain might very well offset psychologically, if not economically, losses to grains.

Having achieved some initial understanding of Nyiha ethnogeography and its meaning within our own scientific system, we can turn to Nyiha agricultural systems, the means by which Nyiha farmers operationalize their system of environmental knowledge as productive activity.

Traditional Agricultural Systems

chapter 4

"Avadumbwe avashilya numbu namahambili."–
Even the rich eat roasted numbu with crusts.
NYIHA PROVERB (Busse 1960:134)

Human societies require energy and material from their environments for survival. At a basic biological level, these needs are sought through other living entities—plants and animals. For most societies, agricultural systems provide the means by which environmental energy and matter are channeled for human use. We often see manifestation of these agricultural systems in patterns written upon the landscape. Underlying these land-use patterns are ordered processes of human activity. This activity includes thought and skills, articulated in words and in actions using tools, domesticated plants, and animals, all set within a particular environmental milieu. Agricultural systems follow the rhythm of the seasons, and although they may change through time, they constitute the critical functional linkage for survival between society and environment. Most, if not all societies seek more than survival. Alternative choices of production are weighed; alternative techniques are selected to cope with survival, varied tastes among foodstuffs, and ultimately, desires beyond simple sustenance.

The Nyiha, as nearly all mankind, sustain themselves through mobilization of their systems of agricultural activity. Understanding of the traditional agricultural practices used by the Nyiha for exploiting the Mbozi environment requires discussion of the Nyiha's complement of tools and crops. Then, we will turn to the land management systems themselves, focusing upon the way in which agricultural practice articulates crops and environment through the seasons and years. Discussion of the role of livestock will provide a transition to our consideration of contemporary Nyiha farming patterns (Chapter 5).

A number of tools and processes are common to most Nyiha agricultural systems. The principal traditional agricultural implements are the long-handled hoe, *ijembe,* and the ax, *intemo,* and machete, *ishipanga,* for cutting trees (Figure 4.1). While wooden ax and hoe handles are still carved locally, hoe blades and pangas are imported from England or Czechoslovakia. Ax blades are most commonly shaped from pieces of motor vehicle spring leaf by heating, hammering, and grinding the blade edge. In

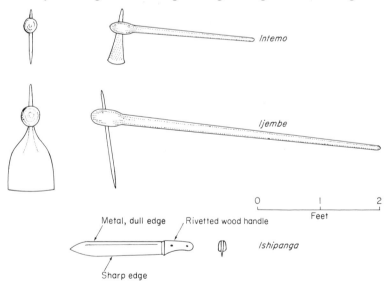

Figure 4.1. Agricultural tools.

recent decades, use of the single-share, ox-drawn plow has become increasingly widespread.

In all Nyiha agricultural systems, most if not all of the following basic procedures are used:

kupemba:	to burn
kuselela:	to hoe the soil
kupesa:	to scatter or sow
kusangula:	to weed
kuvuna:	to harvest
kuaniha:	to spread the crop in the sun to dry

Specific execution of these activities varies from crop to crop and among field types.

Crops

Major staple grain crops of the Nyiha are finger millet, sorghum, and maize. Bulrush millet, a Wanda staple, is known to the Nyiha, but had not been cultivated in the memory of any informant interviewed in 1967. Wheat is becoming increasingly popular, and there have been some local attempts to grow rice. Among root crops, the Livingstone potato, sweet potato, and European potato are cultivated. Cassava is becoming increasingly important as a starchy staple. Taro and yams are known, but not cultivated. Castor was once an important cash crop for the Namwanga and Wanda and is grown by the Nyiha. Among many legumes cultivated, only the common bean, garden pea, and groundnut rate separate fields. All domesticated fruits are recent introductions to Mbozi, except one banana variety which was present in the pre-European period. Red pepper and sugar are the condiments grown, and a large variety of vegetables is cultivated in gardens or interplanted in grain fields. Tobacco is grown by many families and is becoming an important cash crop among the Namwanga of Ndalambo Division.

In the following discussion the major features of the most important food crops grown in Unyiha are summarized (Table 4.1). A few pertinent remarks will also be made concerning crops recently introduced as well as those no longer important in Nyiha agriculture.

Table 4.1

Nyiha Crops

Botanical Name	Common name	Shinyiha	Kiswahili
	Grains		
Zea mays	Maize (corn)	Amangagu	Muhindi
Eleusine coracana	Finger millet	Ulezi	Ulezi
	—early maturing	Inyego	
Sorghum vulgare	Sorghum	Upemba	Mtama
Pennisetum thyphoides	Bulrush millet	Utuele	Mwele
Triticum spp.	Wheat	Ingana	Ngano
Oryza sativa	Rice	–	Mpunga, mchele
	Root Crops		
Coleus esculenta	Livingstone potato	Ululumbu (inumbu)	Viazi maji
Colocasia antiquorum	Taro	Amavindu	Mayugwa
Dioscorea spp.	Yam	Amatugo	Viazi vikuu
Ipomoea batatas	Sweet potato	Impazinga	Viazi vitamu
Manihot esculenta	Cassava, manioc	Mahogo	Muhogo
Solanum tuberosum	European potato	Intofanyi	Viazi vilaya
	Oil Crops		
Ricinus communis	Castor	Impulya	Mbarika, Nyonyo
Sesamum indicum	Sesame	Usambwe	Ufuta
	Legumes		
Arachis hypogaea	Groundnut	Imbalala	Karanga
Cajanus cajan	Pigeon pea	Amawanga	Mbaazi
Phaseolus lunatus	Lima bean	Amankolo	Mfiwi
Phaseolus vulgaris	Common bean	Imponzo	Maharagwe
Pisum sativum	Garden pea	Insyavava	Ngegere
Vicia faba (?)	Broad bean (?)	Mankolovemba	Maharagwe mkubwa

English	Scientific name		
Cow pea	Vigna sinensis	Insonolo	Mkunde
Cow pea (Early)	Vigna unguiculata	Uchezya	Mkunde ndogo
Bambara groundnut	Voandezia subterranea	Inzungu	Njugu mawe

Fruits

Pineapple	Ananas comosus	Ivinananzi	Mnanasi
Pawpaw (Papaya)	Carica papaya	Papai	Mpapai
Lemon	Citrus limon	Masitroni	Mlimau
Tangerine	Citrus reticulata	–	Mchenza
Orange	Citrus sinensis	Nsongwa	Mchungwa
Mango	Mangifera indica	Inyembe	Mwembe
Banana	Musa sapientum	Inkombwe	Mgomba, Ndizi
Guava	Psidium guajava	–	Mpera
Pomegranate	Punica granatum	Komamanga	Komamanga
Tamarind	Tamarindus indica	Amasisi	Mkwaju

Condiments

Red pepper	Capsicum annuum	–	Pilipili hoho
Sugar cane	Saccharum officinarum	Amaguva	Muwa

Vegetables

Onion	Allium cepa	Kanyenze	Kitunguu
Cucumber	Cucumis sativus	Itana	Mtango
Pumpkin	Cucurbita maxima	Iyungu	Mboga
Lettuce	Lactuca sativa	–	Saladi
Gourd	Lagenaria siceraria	Iyengala	Mmumunya
		Urufuru	Mkunga
		Ifuru	Kibuyu
		Ishaji	–
		Inkoloo	–

Other Crops

Tomato	Lycopersicum esculentum	Inyanya	Mnyanya
Native tomato	Solanum sinuato-repandum	Ipwa	Ngogwe
Tobacco	Nicotiana tabacum	Itumba	Tumbako

Finger Millet. Finger millet *(ulezi)* is the major staple in Unyiha. Greenway (1944:180) and Cobley (1956:33) stated that the probable origin of this annual grass was in India where it may have been selected from the wild grass *Eleusine indica*. Murdock, on the other hand, has argued that finger millet was domesticated in highland Ethiopia (1960:525). Whichever might prove to be the ultimate source, it is an important staple in East, Central, and Southern Africa as well as India where finger millet or *ragi* occurs in innumerable varieties (Doraswami 1942; Iyengar 1945–1946). Trapnell (1943:34–35) cited a larger number of techniques of finger millet cultivation in Zambia, but the word stem *-leze* or *-lezi* occurs only as far south as the Fipa–Mambwe and Nyiha–Safwa language groups, which also is a rough approximation of the limit of use of Kiswahili as a *lingua franca*. Nyiha regard finger millet as a traditional staple. Only in the mythological sense of the whole technique of agriculture having been introduced to an aboriginal hunting and gathering people can there be said to exist legends of its introduction from outside Unyiha.

Finger millet is tolerant of a wide range of moisture conditions and yields well even on poor soils, although it is common among the earliest crops in tropical agricultural sequences (Porteres 1951; Webster and Wilson 1966:168). It has a shallow, fibrous root system, but can tolerate a fair degree of drought. In Uganda it is commonly planted before the rains and allowed to germinate with their arrival. If there is a significant break in the rains, however, the crop may fail (McMaster 1962a). In Mbozi, finger millet is not customarily planted until very late November, December, or even January, thus greatly decreasing the risk of a break in the rains during the early period of growth. The growing season is from 5 to 6 months in Mbozi with time allowed for drying on the stalk. An early-maturing variety known as *inyego* is harvested in May, while *ulezi* stands in the field until June or July. Just as finger millet straw is considered good fodder in India (Willis 1909:50), in Mbozi cattle are allowed to graze the stubble of harvested fields.

Finger millet is cultivated in Unyiha on cut and burned fields *(nkomanjila; ntemele* very rarely now) with a yield expectation of 500–900 pounds per acre, based upon estimates made in the field (Chapter 5). Similar yields are obtained from grassland prepared by either the *itindiga* or *nkule* method, while yields from *ihombe* fields are typically 1200–1500 pounds per acre with an extreme of 2000 pounds calculated in one case. The typical Unyiha yields cited, with the exception of those of *ihombe*, are from fields with a small planting of legumes and possibly maize or sorghum intermixed. Finger millet stores well after thorough drying and is easily ground into flour. The crop yields more protein per acre at typical Mbozi yields than the other staple grains, and because of its comparatively greater yields relative to maize or sorghum, caloric yield per acre is also highest of the grains.

Finger millet is consumed in two ways. First and foremost, it is hulled, winnowed, ground, and cooked with water to form a thick mash known as *inzugu* or *uvugali* (Kiswahili: *ugali*). A small portion of *ugali* is picked up with fingers and formed into a cup shape. This is dipped in a sauce, relish, or stew of cooked herbs, beans and/or meat, and eaten. Alternatively, finger millet is the major constituent of beer *(ipele)*. Unhulled grain is put into water and allowed to sprout. Then it is spread in the courtyard to dry in the sun. After drying, it is placed in water to ferment. Beer is consumed within a day or two before it spoils.

Maize. Maize *(amangagu)* is an annual grass of American origin. Most authorities state that maize was brought to Africa by the Portuguese (Miracle 1965). However, there is some suggestion that its introduction to Africa may have been pre-Columbian (Carter 1968; Jeffreys 1953, 1954, 1967). The Nyiha, Safwa, Lambya, and Malila all share the term, *amangagu,* which is similar to other Tanzanian words for maize derived from the word *manga,* referring to Arabia. However, the majority of Corridor peoples have different terms for maize, suggesting diverse paths of diffusion of this crop. Although it is likely the Nyiha received maize through the Arabs or from contacts in Malawi where it is an important staple, Nyiha consider maize a traditional crop.

Maize is the least drought-resistant of the Mbozi crops. It is especially vulnerable to drought during the period of tasseling, 50 to 70 days after planting, and maize planted so as to be vulnerable to a break in the early rains would have yields significantly reduced. Maize is particularly attractive to vermin. In

Mbozi its growing season is 5 months, but it may be harvested early to be roasted on the ear *(amatsantsa)* or be left on the stalk for a month or so past maturity to assure the complete drying necessary for successful storage.

Maize is cultivated in two ways. First, Nyiha use it as an intermixed crop with finger millet and sorghum on fields prepared by a variety of techniques *(nkule, nkomanjila, itindiga)*. In addition, nearly every family has a separate field *(ishizi sha amangagu)* in which maize (rarely, maize intermixed with beans) is planted on ridges *(mandi)* with the intention of harvesting it early for use as a vegetable or for roasting. As an intermixed crop, yields are from 100 to 300 pounds per acre; no yields were obtained for *ishizi*-grown maize which was not customarily put into storage, but simply eaten as needed directly from the field. Young maize stalks are often chewed by people working in the field. The cool, sugary juice is indeed a pleasant refreshment during the warm weather. The grain is made into flour for preparation of *ugali (ishinyanya* when made of maize).

Sorghum. There seems to be no dispute that sorghum *(upemba)* was a product of African domestication (Greenway 1944; Mauny 1953; Cobley 1956; Murdock 1959; Porteres 1962). Greenway (1944:252) cited an early Arab report of sorghum *(dhurra)* as the staple grain in 10th century Zanzibar. Porteres (1959:73–74) suggested that the root *-pemba* refers to the Indian Ocean island by that name. This may indicate an Indian Ocean origin of this particular variety. Sorghum is now an important staple in parts of Zambia (Trapnell 1943). However, in many societies, including the Nyiha, it is a supplement to finger millet and may be a remnant of a more extensive area of cultivation.

Sorghum has great drought resistance. It requires considerably less water during the growing season than does maize (Glover 1948). Its root system is able to tap large areas of surface soil and extends to a depth of 3 to 4 feet. Sorghum has the ability to suspend development in the absence of sufficient moisture. Moreover, while vegetatively resembling maize, it has twice the number of lateral roots at any stage of development; it typically has one-half the leaf area of maize; and the leaves have a waxy coating to reduce water loss, all of which make it drought resistant (Artschwager 1948:6; Cobley 1956:15; Wrigley 1961:78–79).

In Uganda sorghum is grown with finger millet as insurance against drought (McMaster 1962a).

In Mbozi sorghum is sown exclusively in fields with finger millet. The perennial varieties used yield much more heavily in the second year during which they are scarcely tended. Ratoons from subsequent years will be protected if the field is reopened for such hoed crops as beans, but bush fires and weed succession make harvesting of unattended fields in third or later years unprofitable. Almost all of the sorghum planted in Mbozi is white and known simply as *upemba*. This is a highly palatable variety that is resistant to weevil damage in storage. A red variety known as *ububu* is also said to exist, but no specific reasons were offered for its cultivation in place of the white variety. First year sorghum yields while interplanted with finger millet range from 50 to 200 pounds per acre. In the second (ratoon) year, expected yields are from 200 to 600 pounds per acre. Sorghum is nearly always mixed with finger millet in food or beer preparation. What it might yield if Nyiha liked it well enough to attend it closely or plant it separately remains unknown.

Bulrush Millet. Domesticated in Africa, bulrush millet *(uwele)* was formerly widespread in the Corridor region. It requires less rainfall than sorghum and can survive on poor, sandy soils where other grains would fail. It matures in a shorter period than sorghum (as little as 3 months), but is less tolerant of midseason drought. Bulrush millet is extremely attractive to birds, but stores well once harvested (Masefield 1949:30; Cobley 1956:26; Wrigley 1961:78–79; McMaster 1962a:63). The fact that the Nyiha term *uwele* is from Kiswahili substantiates their claim never to have grown the crop. Bulrush millet is a minor staple among the Wanda, a sort of drought insurance crop.

Wheat. Wheat *(ingana)* was first cultivated in Tanzania by Arabs in 1852. Varieties in the Southern Highlands are derived from missionary introductions in the late 19th century. It has become a staple among the Kinga, Wanji, and some Safwa (Greenway 1944:252; Tanganyika Department of Agriculture 1945:34). Only since the 1950s has wheat been grown by the Nyiha. It presently serves both as an alternate staple and a cash crop. It is planted late in the season (March) in order to avoid rust from late rains as it comes to maturity.

Rice. Rice was brought to the Rungwe lakeshore by Arabs in the late 19th century. It is an important cash and staple crop among the Nyakyusa. Several farmers in Nambinzo VDC have tried to grow rice in *mbugas,* attempts that have failed to date. These men had worked for rice farms in Usangu and were attempting to apply knowledge gained there. It is not impossible that broadcast rice might be successful in certain zones of *mbuga* with proper soil preparation and water management.

Cassava. From its introduction from the New World into the Congo Basin by the Portuguese, cassava *(mahogo)* has spread rapidly across tropical Africa (Jones 1957; Jones 1959). The large starchy root is eaten from 9 to 24 months after planting (Figures 4.2 and 4.3). Several varieties are grown in Mbozi (Table 4.2) all of which have arrived since 1900. Three varieties are sweet and may be eaten as a raw snack after peeling. However, these may also be treated like the bitter variety which requires a lengthy soaking to remove the poisonous hydrocyanic acid. It is then dried and ground into flour for *ugali.* The leaves are often used as a relish. Cassava is both drought and locust resistant. While it thrives on heavy rainfall, it will tolerate severe drought and still yield 3 to 5 tons per acre. Cassava in Mbozi seems to be free of the virus diseases problematical in East Africa and elsewhere (Masefield 1949:41–42; Cobley 1956:174–177; Childs 1961; Wrigley 1961:80–81; Purseglove 1968:172–180).

Table 4.2

Cassava Varieties[a]

I. Sweet varieties, safe to eat uncooked although usually cooked

   Ukanyukule long narrow leaves, yellow bark—the preferred variety
   Umwakapila large leaf, reddish bark on branchlets
   Umalia   small leaf, reddish bark on branchlets

II. Bitter variety, must be processed for removal of acid before eating or feeding to livestock

   Untutumsi reddish bark, great proliferation of branches

---

[a]These varieties were named and consistently identified by Nyiha informants.

Figure 4.2. Planting cassava. Cassava is vegetatively propagated from stem cuttings and planted on *mandi* (ridges). Here girls plant cassava with a second crop of beans mid-rainy season. Note the uprooted and detubered bushes from which cuttings are taken.

Figure 4.3. A sprouted cassava cutting. An interplanting of legumes will not only increase the yield from the field but also reduce weeds like the grass at the left.

Inumbu. Inumbu or the Livingstone or Kaffir potato is one of a number of indigenous African species of the genus *Coleus*. This irregular-shaped tuber is apparently indigenous to southwestern Tanzania (Greenway 1944:36–37). *Inumbu* is eaten as a vegetable and is still widely grown in small quantities. The Nyiha proverb at the beginning of this chapter could be interpreted in several ways. My speculation is that *inumbu* was an important insurance crop during the 19th century unrest. During that time stored grain was vulnerable to theft while *inumbu*, as other tubers, could be safely left in the field until consumed.

Sweet Potatoes. Sweet potatoes *(impazinga)* were introduced to Africa by the Portuguese, even though they occurred in Polynesia in pre-Columbian time. They have replaced yams since they yield more for less work; in turn they have been somewhat displaced by cassava. Sweet potatoes are less tolerant of drought than cassava, but require a well-drained soil. Mounds can provide such a tilth. They are vegetatively propagated and are edible in Mbozi after 4 to 6 months, although not harvested until needed. The tuber is usually baked and the leaves used as a pot-herb (Greenway 1944:37; Masefield 1949:40; Cobley 1956:171, 174; Wrigley 1961:80–81; Purseglove 1968:79–88).

European Potatoes. European potatoes *(intofanyi)*, actually a New World crop, were introduced into Tanzania before World War I by German missionaries in Langenburg District (Greenway 1944:37; Purseglove 1968:560–563). They are an important staple in the Poroto Mountains between Lake Nyasa and Mbeya. Nyiha do not like them as well as sweet potatoes, and they are only rarely grown other than for sale to Europeans.

Yam and Taro. Neither yams nor taro have been cultivated within the memory of Nyiha informants. Both crops arrived in Africa from Asia at an early date, and have been displaced in many African areas by other tubers.

The Castor Plant. The castor plant *(impulya)* was raised for the oil *(imbono)* of its seeds and was traditionally used as a body ointment and hair dressing by the Nyiha. It is indigenous to Africa (Purseglove 1968:180–186), and grows semiwild in Mbozi. A large amount of seed was sold through Asian entrepreneurs which eventually reached the world industrial oil market. Today, very little castor is grown in Mbozi. Cosmetic oils are purchased rather than raised.

Sesame. Sesame *(usambwe)*, similarly indigenous to Africa, was used both as a cooking and annointing oil. It is now little grown in Unyiha, and probably never did well in the high-moisture capacity soils of the Mbozi Plateau. However, the crop is an important source of income for the Namwanga of Msangano and for the Wanda in areas where its tolerance of drought is desirable (Hill 1947; Joshi 1961; Purseglove 1968:430–435).

Groundnuts. Groundnuts *(imbalala,* peanuts), a New World crop, were introduced to East Africa by the Portuguese. They have been an important crop in Unyiha primarily for use raw as a snack; ground in *ugali,* in soups, and relishes; but also for their edible cooking oil. With beans, groundnuts constitute an important source of vegetable protein. To provide proper tilth and easy harvesting, heavier soils are mounded for this crop. It is relatively drought-tolerant.Groundnuts are thoroughly dried before storing to prevent rotting (Hartley 1936; Hill 1947:140–146; Cobley 1956:110; Masefield 1949:33–34; McMaster 1962a:77–88; Purseglove 1968:225–236).

Common Beans. Common beans *(imponzo)*, a dwarf variety of haricot, kidney, or French bean, are also a New World crop. Beans are usually dried and cooked as an *ugali*-accompanying sauce or soup, and less frequently eaten green as a vegetable. They can be harvested in as little as 3 months. As a result, two crops of beans can be grown in a field during the rains, or the first beans can be followed by a root crop or wheat. However, toward the drier northwest, rains are not of sufficient duration to grow two crops. They are usually grown on mounds and often interplanted in maize (Greenway 1944:178–179; Masefield 1949:32; Cobley 1956:141-142; McMaster 1962a:83; Purseglove 1968:304–310).

Garden Peas. Garden peas *(insyavava)* are less commonly grown in Unyiha than groundnuts or beans. They were probably introduced to East Africa before the first European contact and are an important pot-herb among many Corridor tribes (Greenway 1944:179; Purseglove 1968:311–315).

Bambara Groundnuts. Bambara groundnuts *(inzungu)* yield a high protein pulse with very little oil. The crop is indigenous to Africa, and many varieties are grown in Zambia. It is tolerant of poor soils and drought and is often grown in areas too arid for other pulses and groundnuts. It may have been grown by the Nyiha at one time; the name *inzungu* suggests that Europeans *(Wazungu)* attempted unsuccessfully at one time to reintroduce it (Greenway 1944:178; Purseglove 1968:329–332).

A number of other pulses are interplanted with grain crops. Among them are pigeon peas (indigenous to Africa), lima beans (a Portuguese or Spanish introduction from the New World), and two varieties of cowpeas (an African domesticate). In addition, the Nyiha crop *mankolovemba*, tentatively identified as the broad bean, is also grown. All of these crops are dried and used as soups for relishes to accompany *ugali*. The effect of the interplanted pulses on grain yields is mentioned later. Intercropping is ubiquitous except in *ihombe* finger millet fields, and there, soil fertility and moisture differences are sufficient to account for higher yields.

Fruits. Of the fruits, only the banana can be considered a traditional crop. While the mango has great antiquity in Asia and Africa, Nyiha informants related its introduction to Mbozi by laborers who traveled to the coast to work, although it is probable that Arabs carried the mango to the Corridor area earlier. Mbozi mangoes are small and fibrous, far from the pleasant fruit typical of mangoes grown in warmer and moister climates. Neither the tamarind nor the pomegranate are common, and they may have been brought to Mbozi by missionaries or settlers. The guava, papaya (Greenway *et al.* 1953), and citrus fruits were missionary introductions to Mbozi, although they were present in Africa much earlier. As an example of the active introduction of new crops, one former Mbozi planter boasted over 50 different fruits on his estate, many of which have yet to be grown by Africans: apples, peaches, avocado pears, various berries, and the like.

Bananas. Bananas *(inkombwe)* were traditionally grown in Unyiha. They are now typical of every Nyiha homestead. Bananas survive the long dry season, but production is seasonal. Bananas were domesticated in Asia, and the route of their introduction to Africa is open to speculation (Murdock 1959:524; McMaster 1962b). Bananas are sometimes harvested in the very unripe stage, dried, and ground into flour for *ugali*. Normally, they ripen on the stalk and are eaten as a fruit. Several varieties are grown, some of which are attributed to Nyakyusa immigrants. Bananas are subject to hail damage and are easily toppled by strong winds, a rarer occurrence (Simmonds 1959; Champion 1963).

Condiments. Red pepper *(pilipili)* was brought to the East African coast by the Portuguese; the crop was probably carried upcountry by Arabs (Greenway 1944:256; Purseglove 1968:524–530). This crop is usually grown in Nyiha household gardens. Sugar cane, an Asian–Polynesian cultigen, was established on the East African coast in pre-European time (Greenway 1944:56). The Nyiha name *amaguva* is etymologically analogous to the Kiswahili *muwa,* suggesting an Arabic dispersion upcountry. The Nyiha never made sugar wine, its only use being as a refreshment eaten in the same manner as maize stalks. The crop is spindly and not grown in any quantity.

Vegetables. The many varieties of gourds, native tomatoes, and pumpkins are traditionally interplanted with grain crops. Gourds were found in America and Africa in pre-Columbian time. In Mbozi they are eaten as a vegetable or allowed to mature and the shell used as a container. Pumpkin is a New World crop of post-Columbian introduction to Africa. The native tomato *(ipwa)* is indigenous to East Africa (Greenway 1944:177–178; Purseglove 1968:119, 124–127). The remaining vegetables are European introductions to Mbozi: onion, cucumber, lettuce, and tomato.

The *Nkomanjila* System

Having reviewed the origin and use of the various food crops, it is now appropriate to turn to the traditional cropping systems. Among these, the *nkomanjila* system has been the most significant in the evolution of the present Mbozi landscape. This system is typical of many shifting cultivation systems, involving cutting of woodland, burning, cropping, and abandonment to fallow regrowth. The locational determinant of *nkomanjila* is, to a large extent, the physiognomic character of the vegetation—*intundu, invuti,* or *indunguti.* Given alternative choices, the Nyiha cultivator prefers *indunguti* or fully regrown or virgin woodland. However, the lack of mature woodland in Unyiha today means that *invuti,* or even *intundu,* will be cut in most cases for *nkomanjila* (Figure 4.4). Within broad vegetation categories, specific trees are used as indicators of areas most suitable:

Amabala:	*Brachystegia utilis*
Ing'anzo:	*B. boehmii*
Ing'anzo:	*B. longifolia*
Ilaji:	*B. spiciformis*
Insani:	*B. bussei*
Ilenje:	*Brachystegia* sp.
Izimbwe:	*Brachystegia* sp.
Intonto:	*Isoberlinia angolensis*
Inchala:	*Acacia macrothyrsa*

Figure 4.4. Cutting bush for *nkomanjila. Intundu* (bush) is cut from June to October. Bush rather than woodland is typical of the densely settled areas near Mbozi Mission.

A dense bush or woodland of other tree species can be cleared if areas of the above trees are not available or not yet ready for clearing. Nyiha recognize that most *nkomanjila* today are made from *intundu*, but they have little choice unless they live in areas like western Wasa or Nambinzo VDC where some *invuti* or *indunguti* is still available.

In burning the *nkomanjila*, cut wood is stacked around tree trunks and burned just before arrival of the rains (Figures 4.5 and 4.6). If a large amount of unburned material remains, it will be gathered and reburned. There is usually a delay of a month between burning and planting. The time in between is used in preparation of other fields. In hoeing *nkomanjila* for planting (Figure 4.7), the ash is spread more evenly through the field, and weed growth is hoed into the soil. Seeds are broadcast and lightly hoed.

The *nkomanjila* must be weeded, usually once, but sometimes twice (Figure 4.8). This, as well as harvesting is the work of women (Figure 4.9), while men have done the cutting, burning, and a portion of the initial hoeing. After harvesting, the crop is spread in the sun to dry before being placed in grainstores (Figure 4.10).

Figure 4.5. *Intundu,* cut and drying. The *nkomanjila* field will be burned before the rains in October or November.

Figure 4.6. The annual firing of cut bush for *nkomanjila* fields. The fires often escape to the surrounding bush and grassland, although grasslands are also fired intentionally to encourage fresh growth for cattle.

Figure 4.7. Hoeing the *nkomanjila* seedbed. A communal work party for field preparation is common, with beer the reward. Finger millet and sorghum will be broadcast and lightly hoed. Commonly, maize and a number of legumes will also be intermixed with these crops.

Figure 4.8. Weeding finger millet. When the millet exceeds 4 to 6 inches in height, it can be distinguished from the related weed, wild finger millet *(Eleusine indica).*

Nkomanjila crops (Figure 4.11) include finger millet; perennial sorghum; a number of pulses, including pigeon peas, lima beans, and cowpeas; and cucurbits including pumpkins and gourds. This system is primarily for production of finger millet, and the other crops are subsidiary to it. Intercropping in this manner serves a number of beneficial functions in comparison to individual fields for each crop (Wrigley 1961; Webster 1966). A crop mixture provides a better coverage of the soil, thereby discouraging weed growth and providing protection against erosion. Varying nutrient and solar radiation demands of crops may make more optimal use of these factors. Varying root depth may similarly better exploit soil moisture resources. An intercropped field may be an insurance policy against pests, diseases, and variable precipitation. Finally, legumes may provide nitrogen for other crops. Two studies of intercropping (Robertson 1941; Evans 1960) both demonstrated that greater

Figure 4.9. Harvesting finger millet. This task is accomplished using a small blade (now, a razor) set into a small handle. This backbreaking task is the women's work.

grain yields result with a small intermixture of legumes. Intercropping from the Nyiha viewpoint, however, must be considered a tradition rather than a practice consciously rationalized.

The standard crop sequence in *nkomanjila* is the finger millet–sorghum complex the first year, followed by sorghum ratoons or suckers the second year. A portion of the first-year field may be planted to *inyego*, an early maturing variety of finger millet. In the second year, the field is referred to as *lisala*. The *lisala* of sorghum ratoons is virtually untended except for harvesting. Traditionally, the cropping sequence would end here and the field would be abandoned to fallow. Today, however, the number of alternative longer crop sequences is extremely large (Figure 4.12). The basic 2-year *nkomanjila* sequence (Line B,

Figure 4.10. Harvested finger millet drying. The crop must be thoroughly dry before storage. If the crop was not sufficiently dry in the field, it is spread (*kuaniha*) on the hard courtyard or wooden racks to dry. This photograph was taken near Msangano, the finger millet having been harvested from an *ichalo,* Namwanga equivalent to the *nkomanjila.*

Figure 4.11. *Nkomanjila* crops. The finger millet, with its characteristic hand-like inflorescence, stands 2 to 3 feet tall. Perennial sorghum is much taller, reaching 10 feet. Local varieties have a compact panicle with white seeds.

Figure 4.13) may be repeated. Alternatively, if the finger millet yield the first year is good to excellent, the basic pattern may be reinitiated in the second (Line F). Other basic sequence building blocks illustrated in Figure 4.12 may be combined as illustrated in Figure 4.13. In many cases, the original *nkomanjila* is subdivided into smaller plots with different sequences as shown in Figure 4.13.

The Nyiha rationale for selection among sequences involves basic considerations that apply not only to *nkomanjila* but other systems as well. First, a supply of starchy staples must be produced. Second, a supply of varieties of staple crops, pulses, and vegetables must also be produced. Third, with certain exceptions (*inumbu* and *ihombe*), when a field produces a successful crop one year, use it again. Fourth, where possible, alternate mounded (*mandi*) and unmounded (*insansila*) sequences. Fifth, if one sequence is doing poorly, try another. Finally, if a field seems tired, either plant a root crop (sweet potatoes, cassava) or

Sequence name	Year 1	Year 2
Nkomanjila	Finger millet, sorghum, legumes, cucurbits	*Lisala*: sorghum ratoons
Beans-finger millet	*Mandi*: beans, maize	*Itindiga* or *infwosa*: finger millet, maize
Beans-maize	*Mandi*: beans	Maize: *Itindiga* or *infwosa*
Cassava	*Mandi*: cassava	*Mandi*: cassava
Sweet potatoes	*Mandi*: sweet potatoes	*Itindiga* or *infwosa*: finger millet, maize
Maize	*Ishizi*: maize	Fallow
Ishizi	*Ishizi*: maize	*Ishizi*: maize
Groundnuts	*Mandi*: groundnuts	*Mandi*: beans, maize
Wheat-beans	Early beans-wheat	Early beans-wheat
Itindiga	*Itindiga* or *infwosa*: finger millet, maize	*Mandi*: beans
Nkule	*Nkule*: finger millet, maize, cucurbits	(Usually goes to another sequence)
Ihombe	Finger millet, maize	(Usually goes to another sequence)

Figure 4.12. Basic Nyiha crop sequence units. "Beans" refers to the common bean, *imponzo*.

Years	1-2	3-4	5-6	7-8	9-10
A	*Nkomanjila*	Fallow ---	-----------	------------	-----------
B	*Nkomanjila*	*Nkomanjila*	Fallow ---	------------	-----------
C	*Nkomanjila*	*Nkomanjila*	Cassava	Fallow ---	-----------
D	*Nkomanjila*	*Nkomanjila*	BF BM M I SP G	Fallow ---	-----------
E	*Nkomanjila*	*Nkomanjila*	BF BM M I SP G	Cassava	Fallow -----
F	*Nkoman* | *Nkomanjila*	BF BM M I SP G	Cassava	Fallow ------------	
G	*Nkomanjila*	BF BM M I SP G	BF BM M I SP G	Fallow ---	-----------
H	*Nkomanjila*	BF BM M I SP G	BF BM M I SP G	Cassava	Cassava
I	*Nkomanjila*	BF BM M I SP G	Fallow	BF BM M I SP G	BF BM M I SP G

Key:

A 2-year crop sequence	BF BM M I SP G	Alternate 2-year sequences: BF: beans-finger millet; BM: beans-maize; M: maize; I: *ishizi*; SP: sweet potatoes; G: groundnuts.

Figure 4.13. Some *nkomanjila* sequences. For the description of crops grown in each 2-year sequence block, refer to Figure 4.12.

fallow it. These constraints are interrelated uniquely within a household production unit and are variable from year to year. Fields will be fallowed sooner when land is readily available. Fields farther from homesteads are more readily fallowed, since older fields require more weeding and should be easily accessible.

Once the *nkomanjila* is abandoned,[1] Nyiha state that it must rest about 5 to 7 years to reach *indunguti* stage; 7 to 12 years to become *invuti;* and 15 to 20 years for *intundu*. The goal of virtually all contemporary Nyiha fallowing for later *nkomanjila* is only *indunguti,* and the Nyiha realize that repeated cultivation and bush fires eventually create *ishihombe,* upland grassland.

[1]Abandonment of the field to fallow may be related to weed growth, lower soil fertility, or a combination of these effects (Nye and Greenland 1960; Ashby and Pfeiffer 1966). The decision to abandon is often based on the higher yields and less labor required to clear a new field rather than weed the old (Morgan 1957; Watters 1960).

Given the population densities achieved in Unyiha in the last several decades, the *nkomanjila* system has been self-defeating. As a system, it is no longer viable since long fallow periods cannot be afforded. As a result of increasingly frequent and prolonged cultivation, a cultivation steppe has been produced in which the agricultural focus has shifted to systems capable of dealing with grassland rather than woodland vegetation. In addition, it is interesting to note that in answer to the open question "How has agriculture changed in the last 50 years (or since your grandfather was living)?" virtually every Nyiha elder stated that larger fields must be made to get the same yields. In essence, there has been a perceptible increase in the area cultivated to provide the food needs of each person.

The *Nkule* System

The *nkule* system is the grassland alternative to *nkomanjila* (Figure 4.14). Techniques used in *nkule* cultivation can be applied both to upland grass communities *(ishihombe)*, resulting in fields known as *nkule*, and to the upper *ihombe* margins, with the resulting fields called *ihombe*. Indicators for the *nkule* method include tall grasses of the *Hyparrhenia* genus, including *igonombila (H. schimperi)*, *impilula (H. variabilis)*, *ivunga (H. collina)*; another *ihombe* grass known as *impilula (Trachypogon spicatus)*; and *inyilu* or *insusu* soils. *Inkanka* soils are generally a negative indicator.

The important feature of the *nkule* system is the mounding of turf and soil over grass (Figure 4.15), with burning of the

Figure 4.14. An *ihombe (mbuga)* margin. On the margin of the *ihombe* the *nkule* method of field preparation will be applied.

Figure 4.15. Hoeing the *nkule* mounds. Grass is gathered in piles and soil hoed on top to form mounds *(mala)*. If too long, the grass may be cut before hoeing. The grass buried beneath the mound is burned. Maize and cucurbits are planted on the mound. It will subsequently be hoed down for broadcast finger millet.

grass under the mound *(mala)*. On the mounds, maize and cucurbits are planted (Figure 4.16). In December, the mounds around these crops are hoed down, the ash and burned soil spread *(kupalasanya)* and finger millet sowed. The field requires two weedings, one in the course of hoeing down the mounds and preparing the seedbed, and a second weeding during the growing season (Figure 4.17). The finger millet crop is harvested and treated as under the *nkomanjila* system. In *ihombe* fields, mounds are made and burned, but finger millet is the only crop planted, after the mounds have been spread.

As woodland has been cut and increasing area converted to grassland, the central Unyiha landscape has taken the appearance shown in Figure 4.18. Large expanses of open upland and *mbuga* have resulted in most fields having been opened since fallowing by the *nkule* method.

Burning of both vegetative matter and surface soil is important in the *nkule* system. In an agricultural system of Shoa Province, Ethiopia, the soil is treated as in *nkule*. There, fallow grassland is plowed during the dry season to break the sod which is then

Figure 4.16. *Nkule* in late December. By this time the mounds have been hoed down and maize and pumpkins are well developed on the former mound sites. Note the details of house construction in the background.

Figure 4.17. *Nkule* in late February. The finger millet is ready for weeding. The fertilizing effect of the burned mounds is evident from the height of the finger millet around the maize stalks, while bare soil can still be seen between the former mound sites.

Figure 4.18. Cultivation of the *ihombe* margin. At Wanishe, upland areas are broken by a number of *ihombe* or *mbuga* grasslands. As upland forest has been cut and as population continues to increase, the importance of *nkule* and *ihombe* fields increases.

hoed into mounds. Cow dung is put on the windward side of the mound and set afire. The soil and turf of the mound is slowly hoed over the burning dung until all vegetative matter has been burned. Huffnagel (1961) who described this system suggested that it was a sacrifice of nitrogen, but that the burning raised soil reaction toward alkalinity (pH increase of 0.2–0.3), making the mineral nutrients present more available. Given the acidity of Mbozi soils, this pH change may be an important function of the grass- and soil-burning *nkule* system.

The important difference between *nkule* and *ihombe* fields is that the *nkule* virtually always initiates a sequence of crops, while the *ihombe* is used for 1 year only (Line A, Figure 4.19) and fallowed a minimum of 3 years. In upland *nkule* fields, the crop sequences are as varied as after *nkomanjila* (Lines B through G, Figure 4.19). Traditionally, the *nkule* system was primarily applied to *ihombe* but as more upland grassland has been created, the method has been successfully applied there as well. An upland *nkule* field is ideally put into a legume–grain rotation (Lines B, C, F, G in Figure 4.19) for 2 to 4 years, then rested 1 or

Years	1	2	3	4	5	6	7	8	9	10
A	*Ihombe*	Fallow ---------		*Ihombe*	Fallow ------------			*Ihombe*	Fallow	
B	*Nkule*	Beans-finger millet		Beans-finger millet		Cassava		Fallow		*Nkule*
C	*Nkule*	Beans-maize		Beans-maize		Wheat-beans	Fallow	*Nkule*	Beans-maize	
D	*Nkule*	Cassava		Cassava		Fallow ---------			Cassava	
E	*Nkule*	Groundnuts		Groundnuts		Fallow		*Nkule*	Beans-maize	
F	*Nkule*	*Itindiga*		Sweet potatoes		Wheat-beans	Fallow	*Nkule*	*Itindiga*	
G	*Nkule*	Wheat-beans		Wheat-beans		Fallow		*Nkule*	Wheat-beans	

Key :   A 2-year crop sequence

Figure 4.19. Some *nkule* sequences. For the description of crops grown in each 2-year sequence block, refer to Figure 4.12.

2 years. Now, many fields of *nkule* origin are cultivated for 6 or more years. As in *nkomanjila,* cassava often ends the cropping sequence, although the wheat–early beans rotation appears to be viable over a long period in Iyula VDC.

Occasionally an *ihombe* field will be planted to finger millet a second year or, if well up on the *mbuga* margin, hoed into large ridges *(mandi)* for beans or groundnuts. The reasons for general lack of success in a second crop of finger millet (or most other crops) may be due to lack of particular micronutrients in *ihombe* fields or to alterations of the soil structure. When the soil is cultivated, an iron accumulation layer begins to harden, causing impeded drainage and waterlogged conditions in subsequent years. A short fallow may reverse this condition.

Subsidiary Systems

Woodland Systems. In addition to *nkomanjila,* Nyiha have another traditional system known as *ntemele* or *itemwa.* In this method, trees cut over a large area are gathered into a smaller circular area *(ivuha)* for burning, identical with the large circle *citemene* of Zambia. Finger millet is planted in the burned area with perennial sorghum. In the second year the whole cut area may be hoed for beans or groundnuts. *Ntemele* is only rarely practiced now. In 1967 I found only one field of this type in Unyiha. This particular field was in western Wasa and was quite small, the cut area being less than 2 acres. However, as late as 1954 *ntemele* practice caused alarm among District Agricultural Officers and the Mbeya District Commissioner. A fine of 100 shillings was imposed for continuing to make *ntemele* (Tanzania Archives File Acc 135 25/1). A fair number of these fields are still made, illegally, by the Namwanga of Ndalambo.

Nyiha informants evaluate the *ntemele* system in the following way. They realize that it appears wasteful of vegetation, but suggest that yields from the small burned area equal the yields that could be obtained were the whole cut area treated as in *nkomanjila*. The real savings, they argue, is in the weeding required on the larger field. They could sow much more densely in the *ntemele* and not weed it. There is a greater opportunity for weeds in the *nkomanjila*, and weeding must be done. Hence, yields in relation to labor from *ntemele* exceed those of *nkomanjila*, and it is this calculation that encouraged the making of *ntemele*. In the eastern hills, *ntemele* was rarely made because the soils are of very high quality.

Grassland Systems. In addition to *nkule* and *ihombe* systems, three additional methods are used in preparation of grassland fields. *Insavi* is used only in Iyula Division and may have been learned from neighboring Safwa, Malila, or Ndali. In this system, tall grass is hoed down and soil turned over part of it in rows. The exposed grass is burned, and the buried grass smolders under the soil. This is done at the beginning of the rainy season, and finger millet is sown immediately after burning rather than later as in *nkule*. The *nkule* must be hoed and weeded before planting as well as during the season, while *insavi* requires the in-season weeding only. The *insavi* system is used on *inyilu* soils with tall grasses, and may reflect a slightly greater reliability of early rains as well as the high moisture-holding capacity of *inyilu* soils in the eastern area.

The second subsidiary grassland system involves turning of sod and planting without burning. This practice is known as *infwosa* when done with the hoe, or *itindiga* (*intindiga, idigima*) when plowed with cattle. Usually cultivation in preparation for sowing is done after the first rains, but it can be done in the

dry season, and if so is known as *insimpa* or *masutu (lusutu,
dust)*. *Itindiga* has now achieved greater importance as an upland
grass technique than *nkule*. Its importance is at least partially
a function of the administration which has actively discouraged
any use of fire in agriculture, as well as the spread of the plow.
It probably developed since introduction of the plow, and was
emulated with the hoe.

The *itindiga* is usually a field of maize and finger millet planted
in an area of fallow *nkomanjila, nkule,* or *insavi*. The basic rotation
is finger millet and maize followed by beans and maize on ridges
(mandi). Crop sequences that follow are any of those that could
be applied to *nkule* (Figure 4.19). *Itindiga* can also be used for
planting wheat in fallow grassland. Often wheat will be preceded
by a crop of first beans planted *insansila* (flat), rather than on
mounds.

The third subsidiary grassland system involves cultivation on
ridges *(mandi)*. Nyiha have traditionally made ridged fields for
cultivation of legumes and root crops. Traditional ridges known
as *ivinondo* were small. Larger ridges, *mandi* (Kiswahili, *matuta)*,
were introduced by Nyakyusa. *Mandi* are constructed from fallow
grassland or old fields by hoeing soil along one elongated strip
on top of the vegetation of an adjacent unhoed strip. The average
mound size today is some 2 to 3 feet high and 6 feet from
ridge to ridge. Partially at the behest of government officials,
the ridges usually follow the contour. Nyiha once hoed mounds
from grassland (known as *intumba*) during the dry season for
cultivation of sweet potatoes. This crop is now grown on *mandi*,
as are beans, peas, groundnuts, and cassava. The *mandi* system
is increasing in importance as a fallow-breaking technique, and
initiation of crop sequences with cassava is increasingly evident.
Mandi are usually constructed after the soil has been wetted
by the first rains.

Ecological functions of ridging or mounding have been cited
by a number of authorities (Morgan 1957; Sauer 1959; Eder 1965;
Webster 1966:132–133, 142–145). Advantages of mounding include:

1. Better aeration of the soil
2. Better soil structure, especially for root crops
3. Easier harvesting of root crops

4. Better drainage and prevention of waterlogging in heavy soils
5. Greater water retention at root depths over one foot in dry conditions
6. An increasing frequency of nitrate flushes due to larger moisture fluctuation in the upper few inches of soil
7. More rapid decomposition of organic matter
8. Concentration of the fertile topsoil
9. Burying of weeds and consequent enrichment of soil by organic matter
10. Viable means of clearing and managing grasslands without the plow
11. Easier weeding
12. Greater soil surface for soak-in
13. Greater rainfall soak-in if water is held back by ridging on the contour
14. Control of erosion

Few Nyiha informants recognize more than one or two of these functions. Ridging as an erosion control measure has been propagated by agricultural extension officers with sufficient success that this factor is well known. More perceptive Nyiha note advantages regarding green manuring, root crop tilth, and ease of weeding. For most Nyiha, the *mandi* system is traditional and appears to give better yields than flat fields—thus it works. That seems to be sufficient.

Maize Systems. The general term *ishizi* or *ishizi sha amangagu* is used for fields of maize or maize interplanted with beans and not ridged *(insansila)*. The phrases *ishizi sha kwisole (ishizi,* far away) *and ishizi sha kuhaya (ishizi,* near home), are common. Usually, the maize from *ishizi* is consumed green rather than allowed to mature fully. *Ishizi* are located adjacent to the homestead where they are fertilized by household refuse or are alternated with beans, or cultivated in valley bottoms as part of a garden. The crop sequences are either continuous maize; alternating maize and fallow; or alternating maize and beans.

Inumbu. Livingstone potatoes are cultivated in fields known as *ishilumbu.* The herb *ulusongole (Latana camara)* and the tree *insugwa (Syzygium guineense)* are positive indicators for *ishilumbu.*

Ishilumbu are used for 1 year only, although volunteer *inumbu* will be harvested in the second year. *Ishilumbu* are usually too small (less than one-tenth acre) to be profitably followed by other crops, although peas or the beans–maize sequences do follow *inumbu* occasionally. Nyiha claim *inumbu* can never be planted twice in succession in a field without an indefinite rest. They attribute this to moles *(utunko)* who are particularly attracted to *inumbu*. In some parts of Iyula VDC, *inumbu* will not succeed even once due to this pest. The herb *itozu (Acalypha spp.)* is a negative indicator, since it is a dominant weed in *ishilumbu* and temporarily remains after the crop has been taken.

Gardens. The traditional Nyiha garden *(vilimbika)* is made by mounding soil in swamps and stream bottom alluvium. Year-round cultivation of vegetables and vegetable maize is possible there if the stream is perennial. Orange and lemon trees are usually planted in the vicinity of *vilimbika* where ground water is available throughout the year. Today, coffee nurseries are also part of the garden.

In the Iyula vicinity a new garden type, *impongolezya*, has been developed in the last few years. Here, larger *mandi* or *insansila* fields are irrigated from upslope ditches. The ditches remove water upstream and carry it at a lower gradient. At a distance downstream, irrigable land between ditch and stream becomes sizable. The ditching was learned from the system used to supply water to coffee-processing machinery on the estates. The verb *kupongola* refers to the opening and closing of the ditch walls in allowing water to irrigate a field. *Impongolezya* in Iyula were being used for maize and bean production in the dry season. The system required cooperation of all *Kumi-kumi* members who had participated in the ditching and making of fields.

Livestock

The extensive *mbuga* areas and grasslands of Mbozi would seem to hold high potential for stock rearing. However, marked climatic seasonality is reflected in decreased palatability and protein content of grasses during the dry season (Whyte 1962). Nevertheless, livestock are important in Unyiha, both as sources of food and labor. Cattle *(ing'ombe)* are the most important large animal; goats *(imbuzi)* and sheep *(ingole)* are kept as well. Every homestead has chickens *(inkuku)* and many also have domesticated doves (pigeons, *inkunda)*. These animals and the dog *(imbwa)* are considered to be traditional; remains of cattle, goats, chickens, and dogs date from the 13th century at nearby Ivuna (Fagan and Yellen 1969). Ducks, donkeys *(indogovi)*, pigs *(inguruwe)*, and cats *(unyavo)* are attributed to European introduction.

Cattle

The cattle of Mbozi are short-horned Zebu (Figure 4.20) ultimately derived from Asia. They have the thoracic hump which is diagnostic of the breed (Joshi *et al.* 1957; Mason and Maule 1960). Epstein (1955) has suggested the use of "chest-humped Zebu" since some varieties of the breed do indeed have large horns. Only the short-horned types are raised in Mbozi, although occasionally long-horned cattle of the Sanga type are sold in cattle markets by Fipa.

Figure 4.20. The typical short-horn Zebu bull of southwestern Tanzania. Milk production is low in the cows of this breed, but the milk appears to have a high butterfat content.

Ehret (1967) has argued on linguistic grounds that cattle spread into eastern and southern Africa before arrival of the Bantu, and that subsequent to the Bantu migrations milking spread southward from north-central Tanzania to Bantu in the southern half of the continent. The early cattle breed may have been the long-horned (neck-humped) Zebu which reached Africa around 1500 B.C. (Webster and Wilson 1966:328) and is represented in southern Africa today by the Afrikander breed. We know from the Ivuna excavations that cattle were present early in Corridor history. Cattle are presently distributed in highland areas throughout the Corridor region (Figure 4.21).

Figure 4.21. Cattle in the Tanganyika–Nyasa Corridor. Areas in which pastoralism is dominant are represented by black, those in which cattle are present are stippled, those in which cattle are absent or rare are clear.

The short-horned Zebu was brought to East Africa by the Arabs in the late 7th century and spread inland along the slave trade routes (Epstein 1955; Webster and Wilson 1966:329). The Mbozi varieties are all colors, and are moderate-to-large framed beasts only rarely attaining 1000 pounds live weight. Like most Zebu breeds, they are slow to mature and tolerant of high temperatures.

In 1906 Fülleborn reported that Nyiha had few cattle, not because they were potentially unimportant, but because of the warfare through the late 19th century. Sheep and goats were plentiful at that time. Nyiha informants told Brock (1968:76) that they received cattle from the Lambya only in the 19th century. Although I did not receive the same information, such a claim may simply reflect loss of cattle during the period of unrest (the earlier breeds?), with replacement later in the century (the later breeds?). It seems unlikely that the Nyiha would have been previously without cattle. By 1926–1927, the cattle population had grown considerably (Tanzania SHPB):

Unyiha: 15,575 cattle
Unamwanga: 5,182 cattle
Uwanda: 1,784 cattle

In 1965 there were nearly 50,000 cattle in Mbozi Area, with the largest numbers concentrated in Unamwanga and Uwanda (Figure 4.22, Table 4.3).

The Mbozi Plateau and surrounding area is generally regarded as being free from the tsetse fly *(Glossina* spp.), the principal vector of *nagana* or livestock sleeping sickness (trypanosomiasis). However, there was a threat of spread of the tsetse from Northern Rhodesia to the southern portion of Mbozi Area in the middle 1930s which did not materialize. In 1967 several cattle died of sleeping sickness on an estate in Mbeya Area, in the area adjacent to Mbozi across the Songwe River. Mbozi itself, though, is too cold for the fly. The low-lying troughs and Rukwa Plain seem to be tsetse-free as well (Knight 1971b).

The major factor explaining the greater concentration of cattle in Unamwanga and Uwanda is the (complete?) absence of East Coast Fever. This fatal disease is caused by a protozoan agent spread by a cattle tick *(Rhipicephalus appendiculatus)*. The disease is distributed in East Africa everywhere the cattle and tick live to-

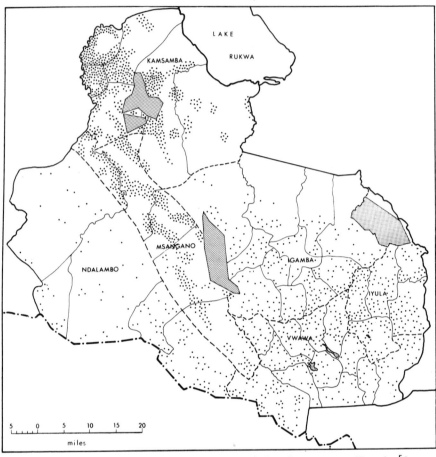

Figure 4.22. Cattle in Mbozi Area, 1965. One dot represents 25 cattle. [Source: Tanzanian Government Agriculture Division, Mbozi Area, Livestock Census. May–August 1965. Manuscript materials on file at the Mbozi Area Agricultural Office.]

gether which includes nearly all the tsetse-free highlands of Tanzania. Mbozi Plateau and the Ufipa Plateau are typical areas of endemic East Coast Fever. Most calves are infected early in life; adult cattle are immune survivors of this early exposure. Nyiha recognize the relative freedom from the tick in the Msangano Trough and Rukwa Basin. There is only a small probability that cattle brought from these areas to central Mbozi will survive. Such transfers occur during the dry season when the tick population is said to be smallest. Nevertheless, survival rates of these cattle are probably no greater than those of locally born calves, as low as 30% in analogous areas in Iringa District (Yeoman 1956:2).

Cattle dipping in Mbozi using arsenical dips is limited to one estate (to which local people may bring cattle for dipping at a break-even fee) and a new dip built along the Great North Road at the Myovisi River crossing. Several other European farmers have a spray system. Since cattle must be dipped about once per week to effectively maintain freedom from ticks, the radius of effective service of a dip is fairly small (under 6 miles). The annual cost of dipping an animal will not exceed 10 to 12 shillings, well worth the investment even in this area of low cattle prices, an adult animal being typically worth 200 shillings or less. In addition to controlling the tick vector of East Coast Fever, dipping reduces mange, cattle lice (*Haemotopinus* spp.), and biting and "house" flies (Yeoman 1956:2).

Anthrax and black quarter are also endemic in Mbozi Area. They can be prevented by annual inoculations which are provided by the Veterinary Department. Cattle offered for sale at the monthly livestock markets at Mbao, Mkulwe, Kamsamba, and Msangano are inspected. Virtually all of the cattle purchases at these markets are for slaughter in Mbozi or Mbeya. If an outbreak of either disease is discovered, the market is closed until the disease can be traced and infected animals destroyed.

In Unyiha, cattle are pastured (*idimilo*) under constant surveil-

Table 4.3

Livestock Distribution[a]

Sub-division	Number of cattle	Percentage composition by livestock units[b]				
		Cattle	Sheep	Goats	Donkeys	Pigs
Iyula	4,866	95.7	1.9	1.8	–	0.5
Igamba	4,212	95.2	3.3	0.9	0.4	0.2
Vwawa	6,482	95.5	2.1	2.2	–	0.1
Msangano	3,085	95.2	0.8	3.0	1.0	–
Ndalambo	11,678	99.4	0.1	0.4	–	0.1
Kamsamba	16,836	98.5	0.2	1.1	0.1	–
Mbozi Area	47,159	97.5	1.0	1.3	01	0.1

[a]Data from: Tanzanian Government Agriculture Division, Mbozi Area. Livestock Census. May–August 1965. Manuscript materials on file at the Mbozi Agricultural Office.

[b]One cow or five sheep, goats, donkeys, pigs = one livestock unit.

lance during the day and placed in a corral *(iwiji; boma,* Kiswahili) or hut *(ishiwaga)* at night (Figures 4.23 and 4.24). Cattle
are the responsibility of males. Young boys herd the cattle (Figure 4.25) and they or men do the milking, although there is no
taboo against women milking. Herd boys make up songs about
cattle, but other than their role in bride-price, cattle are not an
important cultural focus among the Nyiha. There are no rituals
concerning cattle, and they are not groomed or otherwise pam-
pered. Milk *(iziwa)* is usually allowed to sour *(ilizu)* before being
eaten. The entrails *(ulwanda)* and meat of slaughtered animals
are eaten, usually in stews or roasted; and the hides are used for
drums, bed lacings, and rugs. In recent years, the hides have
been stretched to dry and sold through the cooperatives for export.

While cattle slaughter may have been limited to important
occasions in the past, today butchers have established themselves as specialized traders. In remote areas, butchers may
slaughter only one or two animals per week. In the more wealthy
and populous areas or villages, several beasts are killed daily.

Figure 4.23. The *iwiji* or cattle boma. Cattle are placed here every night.
Formerly for protection from wild animals, the *iwiji* now simply prevents the cattle from straying. On the left, a small field of tobacco for home consumption can
be seen.

Figure 4.24. The *ishiwaga* or roofed cattle boma. This type of protection may have been derived from the Nyakyusa practice of stall feeding in similar structures. Predators could not jump into the *ishiwaga*. In addition, this structure provides shade for pregnant cows and calves.

Cattle sales (Table 4.4) are an important source of income, especially in Unamwanga and Uwanda. The total value of Mbozi cattle sales in 1967 was about 1.1 million shillings (156,000 dollars).

Castration and the use of oxen *(inkambaku)*, sometimes even cows or bulls, as work animals are attributed by Nyiha to the missionary and settler. Two animals are usually harnessed to a light plow. One or two beasts are used to pull sleds made from the crotch of a tree (Figure 4.26). These sleds are used to haul mature *(intanga)* to fields (a European-introduced innovation) and bags of crops to the cooperatives for sale.

Minor Stock

Compared to cattle, sheep and goats now have a very minor role in Mbozi (Table 4.5). Sheep are the fat-tailed variety which came to Africa from southern Arabia (Epstein 1954). Presumably

Figure 4.25 Cattle herding in Unyiha. Herding is the responsibility of young boys, traditionally from age 5 through marriage. Now boys herd cattle until a younger sibling can take over so that the older child can attend school.

Table 4.4

Mbozi Area Livestock Sales, 1967[a]

Type	Number	Average price[b]	Total value
Bulls	801	194.60	155,885
Castrates	1231	298.57	367,545
Cows	2366	206.70	489,065
Aged stock	143	118.46	16,940
Male immature	165	93.24	15,385
Female immature	612	115.20	70,505
Total Cattle	5318	209.73	1,115,325

[a]Data are from the Mbozi Area Agricultural Office 1968. No figures are available for other than cattle sales.

[b]Prices shown are in Tanzanian shillings.

Figure 4.26. Cattle and sled. The use of cattle to pull sleds was probably introduced by Moravian missionaries about 1900. Here manure is being hauled to the coffee planting.

the fatty tail gives a greater drought resistance than breeds without this feature. Epstein (1954) attributed the rapidity of the spread of the breed to its supply of fat for people who did not have the pig. In Mbozi, sheep are kept for their meat

Table 4.5

Distribution of Minor Stock in Mbozi, 1965[a]

Subdivision	Sheep	Goats	Donkeys	Pigs
Iyula	490	461	8	125
Igamba	726	212	83	37
Vwawa	717	757	12	32
Msangano	128	490	164	0
Ndalambo	67	208	17	70
Kamsamba	201	970	81	0
Mbozi Area totals	2329	3098	365	264

[a]Data are from Mbozi Area Agricultural Office files.

and fat. They are often herded with cattle or tethered to trees. They are housed at night under grainstores or in special elevated structures (Figure 4.27) which once afforded protection against predators. Mbozi goats are the small, short-eared East African variety (Mason 1960:120). They are kept for their meat and are not milked. They are housed at night in the same manner as sheep.

There are few pigs and donkeys in Mbozi. The one farmer I visited who owned a pig kept it perpetually housed and fed it household scraps, coffee hulls, and wild fruit in season. It was a sow he intended to breed rather than slaughter. Donkeys are used as pack animals.

Chickens abound at every household. The meat is frequently fried or made into stew to accompany *ugali*. Eggs *(idji)* are usually

Figure 4.27. *Ishiwaga* for small stock. Often the bottom portion will be enclosed for calves. The ramp being used by the goat *(imbuzi)* is removed at night and the entrance blocked.

fried, but the Nyiha claim that they are avoided by women and children, similar to other areas of Africa (Simoons 1961). Often friends will exchange chickens, perhaps a remnant of risk sharing from a time when predators were widespread. Chickens are used in traditional rituals and constitute a typical gift for a visitor. They are not well cared for, and hens often lay eggs in fields around the homestead. Occasionally, a broken pot *(ulwinga)* is put on a forked stick above the ground *(ishisanza)* for egg-laying out of reach of dogs. Rhode Island Red roosters are being made available by the agricultural extension personnel for the upgrading of flocks. Ducks are relatively rare. Pigeons *(inkunda)* are provided with a small house on a pole *(ishikunda)* in which to nest. Only the meat is eaten.

Both dogs and cats are scavengers. The most common type of dog is a small traditional breed, but some people have European breeds obtained from planters. The most common of these is the Alsatian (German Shepherd). Dogs are strictly pets, while cats are kept for control of vermin. Many households have dogs; fewer have cats.

Contemporary Agricultural Patterns

chapter 5

"Ijembe wuhalega."–
The hoe will turn against you.
NYIHA PROVERB[1] (Busse 1960:131)

As a casual visitor to Unyiha, you would notice the predominance of farm land—land recently brought into cultivation from fallow for production of staple and subsidiary crops, land allocated to cash crops, and land in various stages of fallow regrowth. Considerable spatial variation in the relative distribution of land uses and field types would stand out, and you would undoubt - edly question why this should occur. Homesteads dot the landscape, with grainstores, buildings, and accoutrements for livestock reflecting the distribution of actual farming units, households. Although no boundaries are marked by fences, people have created both a visible and mental structure on the landscape. Variations in land use impose a grid of organization,

[1]Said when a person has had an exceptional yield one year while his neighbors have not, suggesting that the situation may be reversed in the following year.

and patterns of community consent have created a mental cadastral map. Selecting among the myriad of farmsteads you see, a visit to one would allow reflection on the environmental, social, and economic constraints that create contemporary Nyiha agricultural practice. The Mbozi environment, Nyiha tradition, the external world, and processes of the transition into the money economy and rural modernity, all evolving through time, have shaped the farming patterns we see today.

In this chapter we look at contemporary agricultural patterns, focusing first on a single Nyiha farm. Then, a more abstract analysis of farming system distribution throughout Unyiha will suggest the evolutionary processes of agricultural change during recent decades, including intensified food production as well as cash crops and new technologies. Before embarking on this excursion into farm characteristics, however, it will be profitable to review some general considerations underlying Nyiha farming systems.

At the time of European contact, the term "farm" could not have been applied to Nyiha agricultural holdings. Scattered through space and time, farming activity had a fluid spatial and temporal locus, reflecting the widespread availability of land accessible virtually for the taking. With population growth and allocation of land to cash crops, the fluidity of agricultural activity became increasingly circumscribed, particularly in relation to the dispersed homestead settlement pattern. Still lacking many of the mechanical implements we normally associate with farms, the definite spatial bounding and closely knit scheduling of land use within the Nyiha farming system suggest the transformation of scattered, oscillating agricultural practice into functioning spatial units, farms.

On the Nyiha farm, the production unit is the household, a co-residential group, usually tied by kinship. In the monogamous case, the production unit, household, and resident family are synonymous. In polygynous families, each subhousehold focusing on a wife is semiautonomous, with integration in such day-to-day activity as shared field work and play among the children. Because Nyiha society is strongly male oriented, my interpreter was male and all personal contact with the Nyiha without him was through Kiswahili (spoken by few women), my data on the role of women in agricultural decision making is largely conjectural. Most Nyiha men spoke with authority about their agricultural practice, but many consulted their wives during the course of my queries; in several cases the woman joined her husband as an almost equal respondent. This suggests that it would be eroneous to minimize the role played by women in agricultural decisions. The strength of their contribution to decisions regarding cash crops is certainly less. Cash crops are clearly a man's domain; and until more women become conversant in Kiswahili, only men will be able to interact fully with agricultural extension officers whose advice is important for cash-crop production. Patterns of decision making within the household and respective returns to each person for his work await further investigation. We know, for example, that cash laborers are frequently hired to work on coffee, particularly during picking, although labor parties and exchanges have been carried from the food-crop realm to cash cropping as well. But when a wife labors on her husband's plantation, is she compensated? Are coffee receipts pooled within the household? What claim does the wife have on income from her husband's labor or fields?

In addition to the environmental risks discussed in Chapter 3, the Nyiha farmer faces both in farming activity and daily life economic and social risks which temper his pace of change and, in some cases, limit it through factors beyond his control. In the economic realm, Nyiha have been concerned with prices—prices they receive for their produce and prices they must pay for purchased goods. Coffee prices have fallen; and although the cooperative movement is partially organized to equilibrate short-term price fluctuations, these variations are still felt in Mbozi. Rising prices for various goods and services in the face of declining coffee prices creates obvious economic tension. In addition, increasing bride-price heightens this tension. It is also clear that the question of return to investments is an important aspect of change. Whether we think of the risk discussed among young men of a potentially unsuccessful journey in search of a job or the potential returns to improved cultivation techniques or careful coffee processing, as the Nyiha embark farther upon the sea of money, these kinds of reckoning are crucial, both to the Nyiha and to those who would formulate plans for them. We should note that this kind of calculation

is neither new nor novel. Evidence with respect to Nyiha awareness of returns to labor and land suggests that coping with
the constraints of economic risk constitutes an extension of pre-
existing modes of thought.

We shall also see that social risks affect agricultural practice.
In coffee cultivation, fear of the actions of one's neighbors or
of sorcery *(uvulozi)* initially constrained change. The proverb
at the head of this chapter reflects the community suspicion
that was traditionally accorded to one who stepped markedly
ahead of his neighbors. Brock (1966, personal communication)
observed a cleansing ceremony intended to detect objects placed
in a field by sorcerers to induce crop failure. Acts of envy *(fitina)*
appear in other ways. Brock (1966:23) reported that the first
(1961) innovator to build a mud-brick house in Isansa received
medicine under his door to make him ill, although by 1966 he
had become a local craftsman in house building. Presumptuous
behavior by young people is a part of the same complex of
Nyiha attitudes toward neighborly relationships. Here, Brock
(1966:27) documented the destruction of young coffee trees a
boy of 14 had planted in anticipation of his eventual marriage.
Imitation of one's elders in this way was considered impertinent.
I found considerable resistance to mulching coffee in the most
inaccessible areas for similar sociological reasons, although the
general exercise of social control seemed less compelling in 1967
than Brock found 6 years earlier. This may simply reflect the
continued diffusion of modernization and the commitment to
change from more progressive areas.

With these comments on the milieu of Nyiha agricultural
practice in mind, let us turn to a Nyiha farm as an example
of contemporary farming patterns.

The Silwimba Farm

Laiton Silwimba was born (1939) and raised in Wasa VDC
and attended school through standard four. In 1956 at age 17
he went to South Africa to spend 2 years in the gold mines
of the Witwatersrand. He returned home with sufficient funds
to meet the bride-price of his wife. She was raised in Hanseketwa
VDC and had met him in school. They have one son, born
in 1963, and a daughter born early in 1966. Since returning from

South Africa, Silwimba had traveled several times in search of a job. He sold fish in Zambia for 1 month in 1963, traveled to Iringa looking for a job in 1964 and returned, working in the Lake Rukwa fishery for 1 month.

Silwimba had started a small coffee plantation near his father's home after his return from South Africa. By 1965 it was yielding, but only 2½ *debes* (about 50 pounds) from 1000 trees, which covered about 1½ acres. His yield increased to an appreciable two bags (200 pounds) in 1966, and that year he intended to plant an additional 1000 trees, a task only partially completed. He had learned the benefits of mulching and pruning coffee, and continued these practices. In 1967 Silwimba helped me survey fields and make vegetation samples and identifications. He consented to have his farm surveyed and mapped at that time (Figure 5.1).

Soon after Silwimba returned from South Africa, it was evident in his neighborhood that land was becoming scarce. He recalls

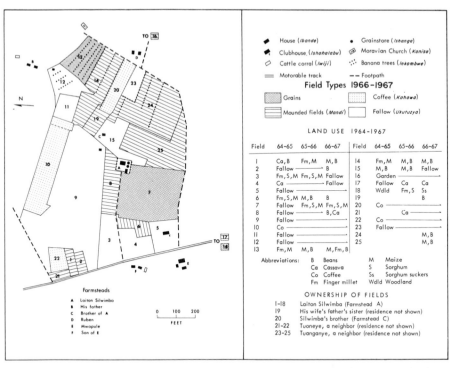

Figure 5.1. A Nyiha farm.

a meeting in the late 1950s with the headman in which the neighbors all agreed to a mutual delimitation of boundaries of land holdings, an enclosure, in fact. The mapped land units shown (Figure 5.1) are those areas alloted to Silwimba at that time. Although he has first priority to use of the land within that boundary, several of the fields are being loaned to relatives and neighbors. A series of disputes involving other minor issues erupted with one of these people in 1966, and he believes the cause underlying the dispute involves the land, lying within the agreed upon boundaries, presently used by the man who is causing the dispute. Most of that land has been used by the neighbor since the late 1950s enclosure.

The 1966–1967 field types and their previous uses are indicated on the farm map. In addition to the crops produced near his farmstead, Silwimba also had two fields several miles to the south in the dissected part of Wasa, one of which was virgin *miombo* until it was cut for *nkomanjila* in 1964. Several hundred yards to the east through Ruben's farm Silwimba has a streamside garden. Ruben is the lay-pastor of the Moravian church.

The majority of the grain fields in production near the farmstead were opened as *itindiga*. For this, Silwimba borrowed oxen and a plow belonging to his father. Silwimba planned to fallow field 7 which was doing poorly and recultivate field 9 as *itindiga* during 1967–1968. Eventually he plans to have coffee throughout field 9, and retrieve the loaned fields for his personal use. The coffee of his younger brother would remain.

Of the 200 shillings (28 dollars) Silwimba earned in 1966–1967 from his coffee, 45 went to taxes and the remainder was spent on clothing, salt, oil for an old lantern, and minor household goods. He, like other Nyiha, had the opportunity to work on a road crew for a month instead of paying the 45 shillings tax, but that year he chose to devote the time to his coffee. He added a kitchen hut to his farmstead in 1967. The clubhouse (*ishahalabu*) belongs to all the men of the neighborhood who drink beer there each week, although Laiton could cultivate the surrounding land if needed.

Like many Nyiha farmers, Silwimba wants to become a major coffee producer. He has seen the material benefits available for cash and hopes someday to purchase cattle, a bicycle, a radio, and a plow. Hence, it is personal drive that carries Silwimba forward. He is too far away from estates to travel for daily work without a bicycle. Land closer to the estates is simply not available. Therefore the only capital impetus available to him is what he, himself, can create. A highly rationalized mixed-farming plan with 5 or 6 acres of coffee could eventually yield an annual income of 300–500 dollars. If this level is achieved, it will result from experimentation, chance, and hard work. Silwimba believes he will succeed.

Agricultural Change

It has been impossible to discuss traditional Nyiha agriculture without reference to change. Nyiha agricultural systems have been evolving from as early a period as can be documented. Here, we will focus on specific aspects of recent agricultural change. Among the changes is an increasing intensity of land use which is reflected in the increasing importance of cassava as a staple crop. A new staple, wheat, is emerging, accompanied by an increasing importance of plows. Among the new cash crops, coffee is the most significant, although pyrethrum has been successfully introduced in the eastern hills of Unyiha. These changes reflect increasing population and desire for material well-being, facilitated by the marketing system.

Analysis of Farming Systems

By turning to detailed interviews and farm surveys conducted in four Nyiha VDCs in 1966–1967, we find an intriguing pattern of spatial and temporal distribution of the basic Nyiha farming systems indicating the on-going process of agricultural change and intensification. In these interviews, a large body of information on field types, crop sequences, yields, and sociological topics was obtained. Wanishe VDC (front end papers) is typical of much of the central Mbozi Plateau. It is relatively densely populated, accessible to both the small settlement at Igamba and to the Great North Road at Mlowo, and served by a coffee cooperative. Wasa VDC represents the transition from the Mbozi Plateau through the dissected plateau margin toward the Msangano Trough. At its western extreme it is relatively isolated.

There, steep slopes and a high percentage of *miombo* woodland are characteristic. Nambinzo VDC represents the arid extreme of Unyiha and is sparsely populated toward the northwest. Iyula VDC is in the well-watered hill land of eastern Unyiha. These site studies provide the basis of analysis of the distribution of modern farming systems in Unyiha

Nambinzo VDC is characterized by a large proportion of fields presently in cultivation as first-year *nkomanjila* (Table 5.1). In nearly two-thirds of all Nambinzo fields, fallow was broken by the *nkomanjila* method (Table 5.2) and nearly all of the *lisala* are the second-year sorghum ratoons. Fields are fallowed more quickly in Nambinzo than in other areas studied, with only 20% of the present fields having been in cultivation for 3 or more years (Figure 5.2a).

In Wasa VDC, *nkomanjila* represents only half the proportion of total fields it did in Nambinzo (Table 5.1). There is a considerably larger proportion of *lisala,* bean, and both first-year and subsequent-crop cassava fields. The *masala* (plural of *lisala*) here are predominantly third- and later-year finger millet fields, rather than sorghum ratoons. Since a slightly larger proportion of all fields were started as *nkomanjila* from the last fallow compared to Nambinzo (Table 5.2), and since 49% of all fields have been in cultivation 3 or more years (Figure 5.2), Wasa seems to have been in a transitional period in 1966. By that time a large proportion of the potential *nkomanjila* land had been cleared, and much of it was staying in cultivation for a long period, eventually to be fallowed and recleared by grassland methods.

In Iyula VDC few of the fields are *nkomanjila* (Table 5.1), whereas *nkule, intindiga, insavi,* and maize are the dominant field types that broke the fallow of fields presently in cultivation (Table 5.2). The high percentage of new grassland fields is reflected in the number of fields in cultivation for only 1 or 2 years, about 56%. Fallows in Iyula under grassland methods are usually short, so that this deceptively large figure does not adequately reflect the intensity of land use there, which is better indicated by the high proportion of grassland cultivation systems.

In Wanishe VDC, the three most common field types were the *lisala* of finger millet and maize, *mandi* beans, and cassava. *Lisala,* bean, and other field types (Table 5.1) that are essentially sequential rather than initial field types total 57% of all fields

Table 5.1

Distribution of Food-Crop Field Types—Unyiha, 1965–1966[a]

Field type	Percentage of fields by study areas			
	Wanishe	Wasa	Nambinzo	Iyula
Nkomanjila	5	13	24	2
Nkule	4	—	1	11
Itindiga	1	4	—	2
Ihombe	9	—	—	—
Lisala (finger millet + maize)	16	12	2	0
Lisala (perennial sorghum)	1	3	11	—
Maize	8	1	6	18
Insavi	—	—	—	4
Beans	14	14	16	13
Cassava	16	25	20	10
Sweet Potatoes	3	5	9	11
Ground nuts	5	8	6	—
Ishilumbu	8	7	5	6
Garden Peas	1	—	—	1
Wheat	5	—	—	11
Vilimbika	4	8	—	4
Impongolezya	—	—	—	7

[a]Figures are the percentage of field types in cultivation in each study area from a farm sample made in 1966–1967.

Table 5.2

Distribution of First-Year Field Types[a]

Original field type	Percentage of fields in use by study areas			
	Wanishe	Wasa	Nambinzo	Iyula
Nkomanjila	10	65	62	3
Nkule	29	—	2	49
Itindiga	19	10	2	12
Ihombe	12	—	—	—
Insavi	—	—	—	12
Sweet potatoes	2	4	10	—
Cassava	17	10	16	6
Maize	2	—	3	14
Ishilumbu	9	5	—	—
Beans	—	6	5	—
Wheat	—	—	—	4

[a]The original field types are the preparation systems used to open the field from its last fallow. Gardens *(vilimbika)* are not included.

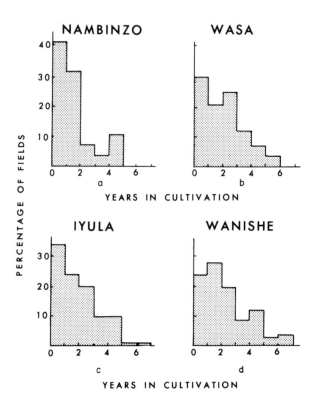

Figure 5.2. Distribution of field longevity. The percentage of fields remaining in cultivation for various periods since last fallow is indicated for four sample villages.

in the farm units sampled. In Wanishe, cassava is often a new field of ridges *(mandi)* hoed from fallow grassland. The remaining new field types are predominantly *ihombe* and *ishilumbu* which are fields customarily used only 1 year, with *nkomanjila, nkule,* and *intindiga* of minor importance. It is not surprising, then, that 50% of the fields in cultivation in Wanishe have not been fallowed for 2 years or more (Figure 5.2d). Few of the fields presently in cultivation were begun as *nkomanjila,* 90% having been derived from one or another of the grassland cultivation methods.

There is an obvious spatial and temporal sequence here. Nambinzo represents the early frontier stage of cultivation with its high proportion of woodland-derived fields. Wasa is in the process of transition to longer periods of cultivation with shorter fallows. Iyula and Wanishe have reached comparable stages where most of the fields are derived from fallow grassland rather than woodland. Wanishe is probably the most advanced in this progression, with the greatest proportion of fields in cultivation more than 2 years.

In terms of an evolutionary model, Nambinzo represents a forest or bush fallow agricultural stage. Wasa is in the transitional period from bush fallow to short fallow, while both Iyula and Wanishe represent short fallow moving toward annual cropping (Boserup 1965). Hence, the agricultural geography of Mbozi in 1967 was not simply a human response to differing resources, but a superb example of the interrelationship between historical process and spatial evolution. Here, the major determinants of this evolution have been the differential population growth within Unyiha and the historical expansion out from the central Mbozi core area.

Agricultural Intensification and Nutrition

In our discussion of the distribution of modern Nyiha agricultural techniques we saw the increased intensity or frequency of cultivation that characterized an on-going process of change. Nambinzo was an example of the early part of this process, and Wanishe and Iyula were the patterns to which Nambinzo would move. An important consequence of this evolution is the Nyiha observation that larger fields are now required to produce an equal amount of crops.

In the past few decades, there has been a notable increase in production of cassava. This increase is partially a function of the locust risk but it is also a response to the decreasing availability of land for starchy staples. It is this argument that should be examined in detail.

Typical Nyiha crop yields (Table 5.3) were determined by surveying fields and obtaining production data through such informant measures as number of *debes*[2] produced, or through measurement and calculation of grainstore capacity to the level filled by the harvest as indicated by the man, often in consultation with his wife. There appears to be a great deal of error in numbers of *debes* produced, which was consistently underestimated by

---

[2]A *debe* is a 4 Imperial-gallon tin, approximately equal to 0.67 cubic feet or 0.54 bushels. A *debe* holds approximately 32–35 pounds of shelled grain (maize, finger millet, sorghum, wheat). For further information on weight equivalents, see Hubbell 1965; Masefield 1949; Wrigley 1961.

a factor of two to five compared with grainstore capacities for the same field and year of production. Measurements based on grainstore capacity seemed on the whole to be more reliable, but must be considered rough estimates rather than accurate data. Sources of error in yield estimation by field and grainstore measurement include:

1. Nonuniformity of grainstore shape (an example is a grainstore square on the bottom but nearly circular at the top)
2. Accuracy of reporting of the level to which the grainstore was filled
3. Intermixture of crops in the grainstore:
   a. crops of different kinds
   b. crops from different fields
4. Accuracy of reporting of field boundaries
5. Accuracy of surveying
6. Amount of crops eaten before being placed in the grainstore
7. Extent of compaction of such crops as finger millet when stored on the head

However, for the purposes of this discussion, approximate yield figures (Table 5.3) are sufficiently accurate.

Average labor requirements for production of typical Nyiha crops (Table 5.4) are suggested from other East African data for areas with similar preparation techniques. These must serve as a reasonable estimate of labor inputs that are likely to be found in Unyiha, since actual measurement of Nyiha labor budgets has yet to be undertaken. On an acreage basis, labor input is not significantly different for most crops, with the exception of sweet potatoes (which require a large time input for harvesting) and sorghum (which requires considerable less attention than other crops, especially in the ratoon years).

The changing emphasis toward cassava can be understood with respect to land and labor inputs required to produce an annual per capita food supply in Unyiha. Typical Nyiha yields for each crop have been converted, using standard measures, into nutrient yields per acre (Table 5.5). Among the starchy staples cassava is clearly the most productive crop in caloric yield. Expressing this caloric yield as the number of cultivated acres required to support one person for 1 year, finger millet

Table 5.3

Typical Crop Yields in Unyiha[a]

	Yields (pounds per acre)	
Crop	Typical yields	Full range of yields
Finger millet	500–900	200–1700
Finger millet, grown in *ihombe*	1200–1500	150–2000
Finger millet, partial yield with maize, or maize and sorghum	200–700	100–1700
Maize, partial yield with finger millet	100–300	80–475
Sorghum, first year partial yield with finger millet	50–200	50–300
Sorghum, ratoons in second year	200–600	100–700
Beans *(imponzo)*	300–500	100–1100
Wheat	300–400	150–600

[a]Data were derived from measurements in the field 1966–1967. Field areas were surveyed and yields calculated from grainstore size or such informant measures as *debes* of threshed or unthreshed grain.

requires some 0.37–0.73 acres per year to supply the staple needs for a person under current yields in Unyiha, while cassava requires only 0.24–0.32 acres per year, assuming a harvest during the second year of production. Hence, when land is scarce, the rationale for growing cassava is obvious. A second dimension of the change to cassava as a starchy staple concerns labor requirements. If the annual acreage requirement for each staple is expressed as an approximate number of man-days of labor required to cultivate that acreage per year (Table 5.5), cassava again is an obvious choice over grains. For every man-day input in cassava production, at least 3 man-days would be required to produce an equivalent calorie supply from finger millet. Given a desire to allocate labor to cash crops or the labor market, such as work on the estates, cassava is again a logical selection over grains to meet the increased demands for time as well as land.

The shift to cassava also reflects a continued Nyiha concern over returns to labor in the face of necessarily increasing concern

Table 5.4

Some Labor Requirements for Crop Production[a]

Crop	Cultivation	Planting	Weeding	Harvest	Total[b]
Cassava	20–25	8–12	12–50	20–40	60–127 (90)
Sweet potatoes	20–25	30	20–26	90–100	160–181
Maize	20–25	8–16	32–48	16	76–105 (90)
Beans	20–25	16–20	30–35	24	90–104
Groundnuts	20–25	9–20	11–45	40–60	80–150
Sorghum	20	6	4	7	40 (40)
Finger millet	20–25	10–20	30–40	20–30	80–115 (90)

[a]The values shown indicate approximate labor inputs measured in 7-hour man-days for production of crops in East Africa (Fuggles-Couchman 1939:396; Wrigley 1961:149; McMaster 1962a:49–50).

[b]Approximations used in further analysis of Nyiha data are in parentheses.

with returns to land. As mentioned earlier, cultivation of the traditional *ntemele* was justified by the desire to economize with respect to labor. The Nyiha similarly recognize the need to produce more food from a decreasing stock of food-crop land available per person. In comparison to the bush-fallow systems, the labor requirements of contemporary grassland management and weeding are appreciated to be higher. Thus cassava helps to offset the evolution toward more labor-intensive systems by requiring less labor for the same caloric harvest.

One of the most important consequences of the shift from the grain staple crops in Mbozi to cassava is the potentially lower quality of the resulting diet (Table 5.5; Bolhuis 1962). As Clark and Haswell (1966:7) state:

> A community at the lowest level of agricultural productivity, living predominantly on cereals, even coarse cereals such as barley, maize, sorghum, or millet, if they have enough calories, will also receive enough protein; though this is not the case with peoples living predominantly on root crops such as cassava, sweet potatoes, yams or taro.

Cassava, unfortunately, has an extremely low proportion of protein compared to carbohydrates. Maize and finger millet have a much more nutritionally favorable protein:carbohydrate ratio. Interestingly, protein production per acre does not vary signifi-

Table 5.5

Nutrient Analysis of Nyiha Crops

Crop	Yield (pounds/acre)	Nutrient yield—pounds/acre[a]			Yield (calories per acre/1000)	Acres per capita per year[b]	Labor Requirements[c]	
		Protein	Fat	Carbohydrate			Man-days per year	Ratio to Cassava
Cassava, fresh	6000	42.0	—	1800	3338	0.32[d]	14.4[e]	1.3
tubers	8000	56.0	—	2400	4450	0.24[d]	10.8[e]	1.0
Maize, dried as	300	28.5	12.0	204	473	1.16	104.4	9.7
whole meal	500	47.5	20.0	340	790	0.69	62.1	5.8
Finger millet	500	40.0	6.5	360	750	0.73	65.7	6.1
	900	72.0	11.7	648	1350	0.41	36.9	3.4
	1000	80.0	13.0	720	1500	0.37	33.0	3.1
Sorghum (perennial, second year)	300	30.0	9.0	210	470	1.16	46.4	4.3
	500	50.0	15.0	350	785	0.70	28.0	2.6
Beans (*Phaseolus vulgaris*)	300	72.0	6.0	144	411	(1.33)		
	500	120.0	10.0	240	685	(0.90)		
Groundnuts	500	130.0	230.0	50	1260	(0.43)		
Sweet potatoes	6000	108.0	42.0	1620	3290	(0.16)		
Bananas	6000	78.0	12.0	1260	2470	(0.22)		

[a]The nutrient productivity of Nyiha crops is calculated from typical yields (Table 5.3) and standard nutrient tables (Nicholls 1961; Raymond 1940–1941).

[b]Assuming a basic daily staple requirement of 1500 calories (Nicholls 1961: 308–319), the land area required to provide the staple food supply for one person each year is calculated.

[c]Approximate and relative labor requirements per year to supply staple needs are calculated using data from Table 5.4.

[d]Harvest at end of second year; yearly average doubled.

[e]Labor per crop, approximately equal to annual labor divided between two years' crops.

141

cantly among the starchy staples. The lower quality of a cassava-based diet can be overcome by increased consumption of protein in other forms, from cassava leaves themselves, pulses, or animal sources. Fortunately, as cassava has become more common in Mbozi, so too has the economy grown with coffee, wheat, and pyrethrum production. Hence, people seem able to afford meat with greater frequency and regularity than formerly. There are no overall figures for the annual slaughter of animals in Mbozi, but there were more than 40 public butchers in Mbozi Area in 1967, two-thirds of them in Unyiha. In addition, a large number of animals are slaughtered privately, suggesting a significant consumption of animal protein. To assess the impact of this potential solution to the nutritional stress posed by cassava, we must eventually ascertain the way in which protein supplies are distributed among and within households. Infant mortality rates provide a general indicator of nutritional and health care status among a people, and the high rates among the Nyiha (Thomas 1972) suggest substantial room for improvement in capitalizing upon available protein sources.

Fish provide an additional potential source of protein, either imported from Lake Rukwa or from *Tilapia* ponds, some 53 of which had been built in Mbozi by 1964. Construction and management of fish ponds has, in most cases, been done under the aegis of the Ministry of Agriculture. Thus far, the Mbozi ponds have only partially served limited family needs, and it is unlikely that they will prove to be viable commercial ventures capable of competing with the Lake Rukwa fisheries in the near future. Fish in ponds and rivers are caught in traps (*umwono*) or are poisoned using *undindini* (*Trichodesma physaloides*).

Wheat

In addition to cassava, wheat is also emerging as an additional staple crop. Although wheat had been produced earlier on the Mbozi estates, it was only in 1956 that African cultivation of wheat began (Figure 5.3). It is now grown in nearly all of Uniyiha (Figure 5.4). In 1967 none of the wheat grown was marketed through cooperatives, although some was undoubtedly traded locally. Wheat has been accepted as a new staple crop, an alternative for finger millet in *ugali*. A few Nyiha buy yeast and make

Figure 5.3. Wheat, introduced by missionaries and planters, has assumed an increasingly important role in Nyiha agriculture. Although yields are only fair, on a cash basis wheat yields equal or surpass finger millet.

bread from wheat which is crushed into a doughy ball similar to traditional *ugali* before being put into the mouth. In central and eastern Unyiha, wheat follows a crop of early unmounded beans, and matures using soil moisture storage lasting into the dry season. It thus draws upon the nitrogen-fixing character of the legume, and the sequence provides both beans and wheat during 1 crop year. Often, wheat is planted by men who first prepare the field with ox-drawn plows.

The Plow

The German administration actively introduced the ox-drawn plow in East Africa. In Mbozi, the plow has been in use since

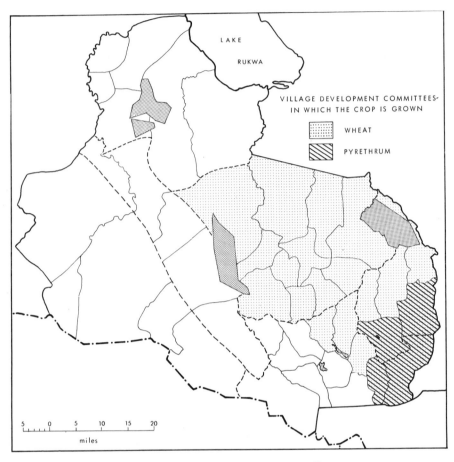

Figure 5.4. Wheat and pyrethrum in Mbozi Area. [Data from Mbozi Economic Survey, Appendix 2.]

the founding of Mbozi Mission. However, it is only since 1950 that the plow has become significantly popular. There are an estimated 5320 plows in Mbozi Area (1967), roughly one for every five to six families. In many Village Development Committees virtually every *Kumi-kumi* group has at least one plow. The plow seems to be making an important contribution in Mbozi toward expanding the land in cultivation and relieving seasonal labor bottlenecks in preparation of fallow grassland for planting. The ox-drawn plow is also extremely effective for cultivating among coffee trees when the oxen are well-trained and the farmer skilled in handling them.[3]

Cash Crops

Sale of African-grown cash crops through the various Mbozi cooperatives in 1967 totaled some 3.7 million shillings (518,280 dollars) in value (Table 5.6). Finger millet was the principal traditional crop produced in excess of home needs for sale—a large proportion of it coming from Unamwanga. Of the cash crops,

Table 5.6

African Cash-Crop Production and Value–Mbozi Area, 1967[a]

Crop	Average price[b]	Tons	Value
Arabica coffee (parchment)	1/- per lb.	1254	2,508,000
Arabica coffee (*buni*)	1/- per lb.	57	114,000
Pyrethrum	2/40 per lb.	66	354,816
Tobacco	2/20 per lb.	2292 (lbs)	5,042
Finger millet	-/50 per kilo	418	468,160
Sesame	-/75 per kilo	113	189,840
Maize	-/22 per kilo	44	21,683
Beans	-/50 per kilo	38	31,360
Castor seed	-/35 per kilo	5	3,920
Rice	-/46 per kilo	2	2,061
Groundnuts	-/70 per kilo	2	3,136

[a]Data are from the Mbozi Area Agricultural Office (1968).
[b]Values are expressed in Tanzanian shillings.

[3]Ruthenberg (1964:183–186) has argued that the ox-drawn plow may not be a viable innovation. Among the questions he raises are the expense; risk of diseases; the supposed necessity to clear the land of tree stumps; and lack of grazing land. In Unyiha, however, people have become adept at plowing around standing fruit trees in fields as well as among coffee trees. The *ihombe* provide land for grazing, as do fallow fields. If disease risk can be overcome, the ox-drawn plow promises to contribute to the further intensification of Nyiha agriculture.

coffee was clearly the most important with sales of 1311 tons worth 2.6 million shillings (367,080 dollars). Other major cash crops are pyrethrum, sesame, and Turkish tobacco.

Mbozi Area fits perfectly the model which suggests that areas isolated from markets by high transportation costs must produce cash crops with sufficiently high value per unit weight to stand the cost of transportation (Clark and Haswell 1966:157). Coffee, which is sent to Moshi in northern Tanzania for processing and marketing, was worth 2000 shillings (280 dollars) per ton in 1967. Pyrethrum in the form of dried flowers was worth two and one-half times that value, as was Turkish tobacco. Virtually all of the other crops will not be transported far but will be sold by the cooperatives in the Southern Highlands area and especially in the Mbeya market.

Coffee

German settlers and missionaries introduced coffee (*Coffea arabica*) into Tanganyika before World War I, but European coffee planters were strongly opposed to its production by Africans. In Kenya, Africans had not been allowed to grow European cash crops. In Tanganyika, settlers argued that African indolence in coffee production would spread disease endangering their own plantations. Ingham (1965:555–556) has suggested they may have been anxious about their future supply of labor as well. Perhaps they also feared theft from their estates. However, this resistence was unsuccessful. The paramount status of the African in Tanganyika was established, and his right to produce cash crops sanctioned in the north under the British administration in the 1920s. There, the Kilimanjaro Native Planters' Association was founded in 1924 by Sir Charles Dundas and was highly successful.

It was only after the arrival of the Mbozi planter in 1927 that coffee was grown outside the Mbozi Mission. Brock (1963:39–40) related a Nyiha story of the first African to grow coffee locally. He was both ridiculed by neighbors for growing other than a food crop and threatened with punishment by Europeans who accused him of stealing the seed. The Mbozi European planters were unsuccessful in resisting African coffee production after the Kilimanjaro precedent had been set. The Nyiha treasury supported supply of seed to African planters, and by 1934 some 100 Nyiha produced 5 tons of coffee. In 1937, the growth of the native coffee industry became self-sustaining. A prewar peak production of 51 tons by 577 growers was achieved in 1940. Nyiha coffee had never been of exceptional quality, and in 1958 it was reckoned to be the poorest in Tanganyika due not to inherent faults in the producing region, but to shoddy grading and processing (Tanganyika MDR 1958). By 1960, however, there had been a spectacular improvement (Tanganyika MbDR 1960); and now Nyiha coffee is generally of good quality.

Coffee seedlings are raised in *vilimbika* with a loose cover of grass on a wooden frame to provide shade. Seedlings are transplanted to the coffee plantation in the middle of the rains, and are protected from the sun by cut ferns until well established (Figure 5.5). Coffee trees flower during the early rains, and plantings are periodically hoed for weed control during the growing season by virtually all Nyiha planters. Coffee berries begin to

Figure 5.5. A new coffee planting. Young coffee raised in nurseries has been transplanted to the permanent field. Here the plants are protected against direct sunlight by ferns *(ulusengaselwa)*. The grass rack covers a small, unirrigated coffee nursery.

ripen in May, and heavy picking begins in June and July, extending into August. Coffee is selectively picked as it ripens to a
cherry-red color (Figure 5.6). There are two basic treatments
for the berry. Some producers have hand-powered coffee pulpers
which they and their neighbors use to express the bean from
the cherry. The cherry is discarded and the beans may be graded
by using running water in a channel to separate them by weight.
The bean is then put into fermenting tanks *(ishivundishililo)* for
several days to remove the mucilage. It is removed, washed,
and dried. Coffee in this form is known as parchment coffee
since the bean is still covered with a thin husk and underlying
silver skin which will be removed in later processing after the
crop has been sold to the cooperative. Alternately, the coffee
is allowed to dry with the pulp still intact and sold in this form.
The latter process results in *buni*, a coffee of considerably lower
quality. *Buni* is also used as a term to describe the lowest quality

Figure 5.6. Picking coffee from a young plantation. This well-maintained planting has begun to yield significantly. Fringing plantings of bananas develop into
windbreaks which are thought by extension officers to decrease coffee water
use.

pulped coffee that has been subjected to the grading procedure.

Nyiha are aware of some of the beneficial functions of mulching, and their coffee plantations are customarily mulched (Figures 5.7 and 5.8). Many informants linked mulching with decreased weed growth as well as the fertilizing effect of the decaying organic matter. The fertilizing effect may be due to enhanced nutrient availability related to pH change or to the addition of organic matter. Mulching is also beneficial in decreasing soil erosion, increasing water infiltration, conserving soil moisture, and minimizing temperature variations (Jacks *et al.* 1955). The latter two factors contribute to an increase in soil nitrate concentration since their beneficial contributions continue

Figure 5.7. Cutting grass for mulching coffee. Grassland, once customarily burned, has assumed a new value among those who do mulch their plantations. Others fear that mulched coffee may be easily burned by enemies. The sickle is a European introduction.

Figure 5.8. Spreading mulch grass in coffee. Mulching and allowing shade trees to remain help conserve soil moisture by lowering evapotranspiration, decreasing rainfall runoff, and decreasing weed use of water. This practice was learned from the estates and is now encouraged by extension officers.

well into or through the dry season (Gilbert 1945). At the present time, mulching with local grasses and banana leaves is the cheapest form of fertilizer. If transportation costs decrease and if dense stands of fallow grass become scarce, purchase of chemical fertilizers may become the most economical practice if labor used presently in mulching can be applied profitably elsewhere. However, many Nyiha farmers are still reluctant to mulch their coffee. The main reason for this is the fear that an enemy can destroy the plantation by burning the grass. When I suggested to one farmer in Nambinzo that he could ease soil moisture problems he faced by mulching, he agreed, but argued that he dare not mulch for fear of fire. Indiscriminant grassland burning is also a danger; for this reason the estates maintain open protective fire breaks around mulched plantations.

An increasing number of Nyiha coffee growers are using more elaborate production methods. Many prune their trees, having been taught on the estates by the agricultural extension officers or by a neighbor. A smaller number use manure and spray equipment. Although many are aware of the economic benefits of these practices, they will only be able to institute them when sufficient capital is available.

Attempts have been made to encourage use of central pulping facilities, but these have served only a very localized area. Frequency of trips with the heavy coffee cherry decreases rapidly with distance (Figure 5.9). Alternately, coffee is picked and carried to the pulper at less frequent intervals, with the result being suboptimal timing of picking and lower coffee quality.

Most Nyiha coffee plantations are less than 3 acres in size and are managed by the household without outside help. During the picking season, households may exchange labor or pickers may be hired. The coffee is hauled to the local cooperative, often by bicycle if only a small quantity, or by ox-drawn sleds. The cooperatives check the quality, make a small initial payment, and pool the coffee for shipment by East African Railways and Harbours trucks to the processing facilities at Moshi in northern Tanzania. Nyiha coffee production rose from 495 tons of parchment and *buni* in 1965 to 1420 tons in 1966, falling slightly in 1967 to 1311 tons. This small decline is not alarming since coffee is a biennial bearer and exceptionally good yields are often followed by a mediocre harvest.

Obviously there has been a very heavy investment in Mbozi for coffee production. Yet, there are difficulties in potential further expansion of the crop. Coffee is vulnerable to pests and diseases which can be controlled with investments in control chemicals and tools for their application. Coffee berry disease has not yet reached Mbozi; if import of seedlings from the north continues to be prohibited, Mbozi *may* remain free of this disease. Coffee is also vulnerable to years of low rainfall, for which mulching is the only generally applicable response. While the estates have begun to rely heavily on irrigation, the small holder will not be able to follow. He cannot afford the equipment; and more importantly, the water resources are simply not sufficient. Finally, the Nyiha coffee grower is vulnerable to the world coffee market and Tanzanian production quotas. For the foreseeable future, the possibility for expanding coffee production is circumscribed.[4]

[4]For further discussion of coffee production see Wakefield 1933; Haarer 1956; Robinson 1961, 1962a, 1962b, 1964.

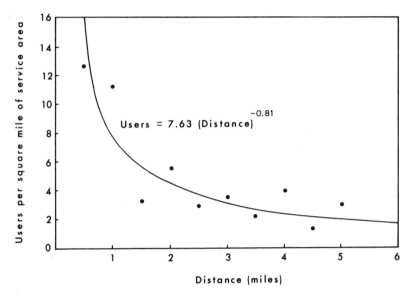

Figure 5.9. Use of a centralized coffee-hulling machine. Data gathered by extension officers in 1963 were converted to pulper users per square mile of service area in rings of one-half mile increments in radius. Note the sharp decline in use with distance.

Other Cash Crops

Tobacco. Tobacco is widely grown by Nyiha and Wanda for home consumption. Among the Namwanga of Ndalambo Division, aromatic Turkish tobacco was introduced as a cash crop in 1964. In 1965, there were 220 growers of the crop, producing nearly a ton of air-cured tobacco. Cultivation and processing of aromatic tobacco requires a fair degree of skill (Driessen 1966); it has thus received a great deal of attention from agricultural extension officers. Today, tobacco in Ndalambo remains a cash crop added to traditional agriculture, and it will be some time before it brings any fundamental transformation of the local economy.

Sesame. A cash crop in Msangano and Kamsamba, sesame is similarly an addition to, rather than a transformation of, traditional Namwanga and Wanda agriculture. Marketed sesame was worth 26,600 dollars in 1967.

Pyrethrum. Pyrethrum (*Chrysanthemum cineriaefolium*) is an herbaceous perennial plant whose flowers contain chemicals known as pyrethrins which are toxic to many insects. In contrast to the commercial chlorinated hydrocarbon insecticides, pyrethrins engender no development of immunities within an insect population and are not poisonous to mammals. The pyrethrum plant (*maua*, Kiswahili for flowers) requires well-drained soils of high fertility and moisture-holding capacity. Rainfall in excess of 50 inches and chilly temperatures to initiate bud development are also required. In the latter respect, even the eastern hills of Mbozi where pyrethrum is grown (Figure 5.3) are at the lower altitudinal limit, and a lower pyrethrin content is found in the flowers there than crops raised at higher elevations. Sixty-six tons of pyrethrum were produced in eastern Unyiha in 1967 (Table 5.6). There was justifiable alarm in both Mbozi and adjacent Mbeya Area over long delays in shipping pyrethrum. Pyrethrin content of the flowers decreases until they are processed, so that their value similarly decreases with shipping delays. The importance of pyrethrum may increase as worldwide use of such persistent pesticides as DDT is increasingly regulated.

The Marketing System

The basis for the modern marketing system in Mbozi is the Cooperative Societies Ordinance of 1932, amended in 1960. The Unyiha Cooperative Society, located in Vwawa, was founded in 1947. Since that time, member cooperatives have been started in Igamba, Msia, Halungu, Itaka, and Ruanda. There are also cooperatives at Kamsamba, Msangano, and Ndalambo. The primary function of the cooperative is the purchase and marketing of crops either locally or through central marketing boards. Cooperatives have a legal monopoly as the sole agents for producers. Sales outside of Mbozi are restricted to the cooperatives, while local sales often take place on an individual basis.

The cooperatives also serve as banks for members and as neighborhood agents for sales of agricultural inputs such as copper spray for coffee, or fertilizer. In the absence of branch banks except at Vwawa (which opened only twice monthly) the cooperative function here is a positive one. In addition, the cooperatives are a medium of information exchange, especially regarding quality control of local coffee processing.

One of the most important inducements to agricultural

development is the certainty and immediacy of farm income (de Wilde 1967, 1:60). The cooperative structure in Mbozi has failed in this respect in two ways. Although small cash advances were paid for crops delivered to the cooperative, final payments for crops were extremely tardy in arrival. During the 1967 coffee harvest, some farmers in Mbozi were receiving the final payments on the 1966 crop. A more extreme example occurred at Kamsamba where farmers highly desirous of starting cash-crop production were easily persuaded to grow kenaf, a fiber crop (Figure 5.10). In 1967, the first kenaf harvest of 1965 had yet to be paid for, and supplies of free seed for a subsequent planting went begging. A second local criticism of the cooperatives was the low prices paid for coffee compared to prices the estates received for coffee of similar quality, some one-half to two-thirds.

Figure 5.10. Kenaf at Kamsamba. Kenaf *(Hibiscus cannabinus)* was introduced as a potential cash crop. Kenaf fiber is stronger than jute and is used for coarse sacking.

The logic of cooperatives having greater administrative costs to handle smallholder production than associations of estate owners and the logic of building cash funds to balance short-term price fluctuations is less compelling than funds immediately in hand. Suspicions of dishonesty are, of course, not unknown. The obvious solution to the price dilemma, sale of the coffee cherry to the estates for processing and marketing, is illegal.

Persistence of the Traditional Food-Crop System

In spite of vast changes that have taken place in Nyiha agriculture, traditional food-crop systems have remained intact. Cash cropping has not revolutionized the farming economy. While there has been a change in emphasis among largely traditional farming systems, all Nyiha families are still self-supporting in major foodstuffs. The modern emphasis on grassland agricultural systems has not altered the family's dependence on its own production for much of what it eats. While the outward signs of a modern, evolving cash-farming economy are emerging, the Nyiha have preserved, perhaps wisely, their traditional food economy. However, cash cropping is becoming more than a veneer, as increasing dependence on market goods grows common and expansion of desires and expectations continues. An examination of the agents and processes of change to date will lend insight allowing speculation regarding the years to come.

Sources of Change

chapter 6

"Ntangavujila kuvubena."–
I cannot return to my childhood.
NYIHA PROVERB (Busse 1960:131)

In Unyiha we have seen the contribution of population growth and cash cropping in the evolution of traditional cropping systems toward the direction of those used for coping with shorter fallow periods and a man-modified landscape. Evenually this process of change will be modeled as a means to discern salient regularities within it; to accomplish this, we must first turn to the sources of change, both those supporting intensified production of food crops and those contributing to the market economy transition. It would be erroneous to think of change as solely contemporary, as we have seen in discussion of the origin of Nyiha crops. Thus, we look first to African sources of change, then turn to alien sources: the missionary, planter, and trader. Finally, the government through *bwana shamba*, the agricultural extension officer, is assessed as a purposeful agent of change. Having acquired familiarity with these sources of change and some insight into the processes by which change emanated from them to the Nyiha, we can then turn to formal modeling of the change process in Mbozi.

The Indigenous Sources

Indigenous African sources have contributed significant agricultural change to the Nyiha. Both migration and intercultural marriages may have carried agricultural techniques from group to group, a process tempered by linguistic barriers in this area where the Kiswahili *lingua franca* has only recently penetrated. In one sense, the population pressure that has caused change might be attributed to alien penetration. On the other hand, it is logical to view the alien influence as accelerating a growth process that would eventually have emerged locally. This population growth resulted in a changing emphasis on existing traditional African practices whose specific origins are discussed below.

Neighboring Agricultural Practices

Documentation of cultivation practices of neighboring peoples in the Nyasa–Tanganyika Corridor provides a basis for discussion of change in traditional Nyiha agriculture. A modest amount of literature exists on agriculture in this region, and for the Tanzanian area, first-hand field observations were made among several neighbors of the Nyiha—the Namwanga, Wanda, Lambya, Malila, Songwe, and Safwa.

Zambia and Malawi.[1] The most common agricultural form of northeastern Zambia is the *citemene* system with its many variations. Basic to all versions of the system are

1. Clearing of large areas
2. Gathering of cut vegetation for burning
3. Making of new fields each year

[1]Trapnell's classic study, *The Soils, Vegetation, and Agriculture of Northeastern Rhodesia* (1943), documents agricultural practice in the present area of Zambia. Miracle (1967) has summarized and made accessible much of Trapnell's work.

4. Cultivation of a set of gardens for each wife
5. Use of rotational sequences
6. Cultivation of finger millet and sorghum together
7. Covering the seed by throwing soil over it with a hoe

In addition, the practice of mounding a field one year and breaking down the mounds the following year is widespread (Trapnell 1943:56). The basic crops of the Zambian systems (in probable order of their local antiquity) are sorghum, finger millet, maize, and cassava. Finger millet has replaced sorghum as the major staple in most areas; similarly, it has been replaced by maize in the Northern Mountain and Eastern Plateau agricultural systems which extend into Malawi. As in Mbozi, cassava is increasing in importance among many peoples. Moffat (1932) suggested that the major advantage of the *citemene* system was that it was the easiest method of finger millet production. Each stage (cutting, burning, planting, fencing) was done "in a burst of enthusiasm and then there follows a long rest [Moffat 1932:60]." Heavy work is done after the harvest when food is plentiful. Other advantages of the system include fertilization by ash and the greater friability of relatively poor soils after burning. Common consequences of the system, however, are the constant movement of villages and apparent widespread destruction of woodland.

Trapnell's Southern Citemene (Figure 6.1) system seems to be indigenous to the Congo-Zambezi watershed. Here a single finger millet crop is planted in tiny circles fertilized by the burning of cut wood. The field is usually hoed for 1 year only. Sorghum is also planted by this small-circle method, as well as by a method known locally as *nkule* in which hoed grass between wood piles is shaken to remove the soil and piled against the wood for burning. Sorghum is planted over the hoed area, while finger millet is planted only in the burned area. The Lala are the most well-known society using the Southern Citemene system (Peters 1950).

Trapnell grouped a number of *citemene* variants together under the rubric, Northern Citemene system. In these systems the cleared areas are smaller than in the Southern Citemene system, one large circle is made, and a succession of crops are cultivated before fallowing. The Bemba are the most familiar society cultivating *citemene* fields, but tradition suggests that they were hoe cultivators while still in the Congo and acquired *citemene* techniques from the Bisa, Lungu, Mambwe, Namwanga, and Wiwa after their arrival in Zambia. Their agricultural system was described by Richards (1939). In the basic Northern Citemene system finger millet is planted in the burned circle only, sometimes interplanted with sorghum, while the remainder of the cleared area is uncropped. A number of crop sequences follow the first-year field, including crops of groundnuts, beans, cowpeas, sweet potatoes, or cassava (Miracle 1967:120–121). The Mambwe, Namwanga, and Wiwa often close the cropping sequence with finger millet on hoed-down mounds, a method called the Developed System by Trapnell (1943). The Masaka variant of the system includes sorghum on mounds as a second- or third-year crop. Among many peoples, if the last crop before fallowing was finger millet, a catch-crop of beans planted on mounds known as *intumba* initiates the following cropping sequence.

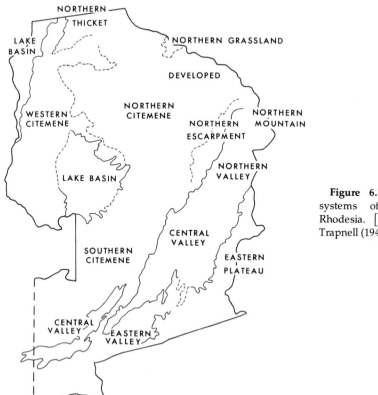

Figure 6.1. Agricultural systems of northeastern Rhodesia. [Adapted from Trapnell (1943).]

Trapnell also described a subsidiary system that occurs among Northern Citemene cultivators, the *nkule* finger millet garden. This system is identical to the Nyiha *nkule* treatment of *mbuga* margins and, in fact, is attributed by the Wiwa to the "Nyika of Tanganyika Territory" (Trapnell 1943:53).

The Northern Grassland system is similar to the *nkule* system, except that the grass is not burned, and a catch-crop of beans is planted the first year. The mounds are known as *intumba*, and when trees occur in the upland grass they are burned before mounding the field. The second year, mounds are hoed down when maize planted in the center of them is about 2 feet high, and finger millet is broadcast as in *nkule* in the first year. This system is dominant among the northern Mambwe as well as the Fipa (Miracle 1967:134–137). Trapnell suggested a Tanganyika origin, and variations of this mounding system do occur into the Lake Victoria region (Rounce 1949). Since there is no fire to destroy weeds, one of the major causes of land abandonment under this system is the pernicious weed *Eleusine indica* (Trapnell 1943:56).[2]

The Western Citemene, Lake Basin, and Northern Thicket systems all have cassava as a dominant or codominant staple with finger millet and sorghum. They represent the eastward migration of the mounded cassava tradition of the Congo Basin (Trapnell 1943:58–64; Jones 1957).

The Northern Valley system uses a method of grassland finger millet cultivation Trapnell called the grass manure method. In its simplest form, grass is cut, seed broadcast, and seed and grass hoed lightly into the top soil. This system occurs among the Zambian Lambya, Fungwe, Tambo, and northern Senga. It is quite similar to the Nyiha *itindiga* or *insavi* systems.

The Eastern Plateau and Valley systems are characterized by the small mound maize culture *(magadi,* Ngoni; *marongo,* (Tumbuka)* in which maize on mounds is weeded, and new mounds formed in the process are used for production in the subsequent year. This system is derived from the Tumbuka and Nyanja-speaking peoples of Malawi and extends into Zambia with the Yombe, who also have *citemene* gardens (Trapnell 1943:79–81). The fields are first prepared during the rains, and

[2]Mambwe agriculture has been discussed in detail by Moffat (1932) and Watson (1958).

beans or cowpeas initiate the cropping sequence. Cereals are planted the following year, and after several years the whole field is devoted to cassava or pigeon peas. This system minimizes land preparation in the hot season before the rains; recognizes that cereals do not do well on unburned, newly cleared land; and utilizes the manure value of weeds (Webster 1966:167–68). The large circle *citemene* also occurs among the Tumbuka where it is known locally as *chisoso* (Wilson 1941). According to Wilson, the *chisoso* method is used where the soil or vegetation indicators cast doubt on the outcome of maize cultivation. *Marongo* cultivation is reserved for fertile soils with good stands of *Hyparrhenia* grass or *Brachystegia* trees (Wilson 1941).

The group of techniques known as the Northern Contact systems includes the Northern Escarpment and Northern Mountain systems. In the former, the Tambo, Lambya, and probably the Wandya make fields known as *nkomangila* (Trapnell's spelling) by cutting *Brachystegia* woodland at breast height. The cuttings are gathered and burned around stumps and the remainder of the field hoed. Finger millet is broadcast with perennial sorghum. Interplanted maize is often eaten green. In the Mountain system of the Nyiha, Wenya, and hill Fungwe *nkomangila* fields are made in *Brachystegia* while *Acacia* sites along streams are prepared by the grass manure method. Ridges in the latter technique are formed by dirt heaped over hoed grass and are planted with maize and beans. According to Trapnell (1943:78–79), ridging was derived from Tanganyika. *Citemene* finger millet gardens also occur. Both Northern Contact systems are analogous to the Nyiha system in Mbozi. The lack of the *nkule* system may simply be due to the absence of the *mbuga* ecosystem in the Northern Contact area. The Central Valley system of the Bisa is essentially the *nkomangila* system with sorghum the dominant staple.

Trapnell created a set of hypothesized relationships he believed to exist among the agricultural systems of northeastern Zambia (Figure 6.2). There are several basic traditions impinging on the area, the Millet Ash Cultures (Small and Large Circle methods), indigenous or an early arrival in the area; the Millet Grass Cultures, possibly derived from northern East Africa; the Small Mound Maize Culture of Malawi extraction; the Hoe-planted Kaffir Corn (sorghum) Cultures, perhaps of greatest antiquity; and the Mounded Cassava Culture, the most recent arrival. The position of the Tanzanian peoples in this scheme will be discussed after a description of systems occurring there.

Tanzania. The principal crops of the Fipa of Tanzania are finger millet, maize, beans, sweet potatoes, and cassava. Finger millet (*amalesi*) is made into *ugali (insima)* as a staple food. The basic Fipa agricultural system of *intumba* mounds has been described above. The 2- to 3-feet high mounds are made at the end of the rains by turning sod on top of grass. A catch-crop of beans is planted, or the mounds may be left to rot until the following rains. The mounds are weeded before the rains, and when the rains begin, hoed down and the staple crops sown. Often cassava planted at the edge of the mound is left when the mound is broken. The crops are harvested in June and July. In the third season, small mounds incorporating weeds and crop residues are made, then broken down in January for finger millet. In the fourth year the field is ridged for maize, beans, and ground-nuts. Ridging is repeated until the soil is exhausted. When trees or bushes are encountered in the initial clearing, they are piled on top of a mound and burned. A small ridge around the bottom of the mound collects ashes washed off by the rain. Here, cucurbits are planted (Lunan 1950; Willis 1966:23). The northern Mambwe grassland agriculture is virtually identical to the Fipa system (Willis 1966:50).

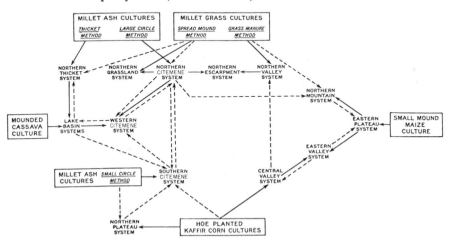

Figure 6.2. Probable origins and development of agricultural systems in north-eastern Rhodesia. [Adapted from Trapnell (1943).]

The *intumba* mounding and *citemene* (known locally as *ntemele*) systems are found in the Ndalambo area of Unamwanga (Figures 6.3, 6.4, and 6.5). The *intumba* is made in an identical manner to the Fipa, and the *ntemele*, as might be expected, is the same as that system among the Nyiha. In Ndalambo the *ntemele* is planted to finger millet and sorghum the first year. Sorghum ratoons are harvested in the second year and the field then abandoned. Namwanga also have the *nkule* method of burning grass on the *mbuga* (Kinamwanga: *matipa*). A few Namwanga plow with cattle in the technique known as *itindiga* in Unyiha. For this system they use the Kiswahili term *inseisa* (Willis 1967).

The Msangano Namwanga also practice the *ntemele* system. Here the first-year field is known as *alipya* (new) and the sorghum ratoon field, *izuka* (old). The alternative to *ntemele* is *ichalo*, a term which simply means field. Here, the tree cover is less dense. Trees are cut and burned along with the grass. The whole area is hoed or plowed for finger millet, sorghum, and interplanted crops. The same crops may be repeated a second or third year, in which case the field is called *ampepe*. The fallow period is known as *alisala*. In recent years some rice has been

Figure 6.3. An *ntemele* field in Unamwanga. This technique is identical to the *citemene* of Zambia. Trees cut from a wide area are gathered into a circle to be burned. Only the burned area is planted. Note the lopping of trees and the light color of dry finger millet in July.

grown in flooded *mbuga* areas. However, neither the *nkule* nor *intumba* methods are used in the Msangano area.

The Wanda make *ntemele* which they call by that name, ridges *(miyandi)*, and *nkomanjila*-type fields called *ivyalo* (Figure 6.6). The staple crops finger millet *(amalezi)*, sorghum *(ampembe)*, bulrush millet *(imbembe)*, and maize *(amakonde)* are often interplanted on new fields *(insinde)*. *Intumba* mounds are only rarely made for sweet potatoes.

A system of cultivation closely akin to both the Fipa mounding system and the *nkule* and *insavi* systems was described in the Mbeya District, presumably among the Safwa (Robertson and Rombulow-Pearse 1955). In this system *Hyparrhenia* grassland is cut after the rains begin. The grass is laid in rows on the contour and one or one and one-half feet of soil heaped on them. A crop of beans or peas is grown on the ridges the first season. In the second season the ridges *(injeleha)* are spread for finger millet, while in the third and subsequent seasons, mounds are again formed for the cultivation of maize. Cassava

Figure 6.4. An *ntemele* ready for harvest. Note the coppice on lopped trees near this field in the Ndalambo area of Unamwanga.

Figure 6.5. The Namwanga *intumba* system. First-year mounds are hoed between trees and planted with beans. During the next dry season trees are cut, gathered around stumps, and burned. The mounds are hoed down and finger millet scattered over the entire hoed area.

is interplanted in the maize as the field's fertility begins to falter and is the last crop planted before fallowing. Supposedly this system has only been used since about 1935 and was not indigenous. It is likely that it is a derivation from the traditional *nkule* system of treating grass plus a large proportion of the widespread *matuta* ridging system.

Safwa east of Mbeya practice the *nkule* mounding system with burning the first year. They also claim to be aware of the *ntemele* system (which they call *ivyala*), and the *nkomanjila* system (sometimes called *nkule* or *ivyala*) is still used in wooded areas. The most common technique is the ridging system described, known as *magunda* (fields) with alternating ridges *(injeleha)* and flat *(ivundiha)* cultivation. The southern Safwa of the Uporoto Mountains have the *nkule* mounding method (called *amakuha*), *injeleha* ridges, and *intindiga*-type fields called *ivizi* which are plowed or hoed after the grass is burned. The basic traditional Safwa agricultural practices seem to be the *nkule* and *nkomanjila*

Figure 6.6. A Wanda *ntemele*. The crop in the foreground is finger millet and in the rear, bulrush millet *(uwele)*. The field is prepared in the same way as the *ntemele* or *citemene*.

methods, to which many variations have been added in recent years.

Songwe agricultural practices are nearly identical to the Nyiha. The *nkule*, *nkomanjila* (here, *ihomanjila*), *itindiga (intinde)*, *impongolezya*, and *ishizi* of maize are common. As might be expected, most crop names and soil and vegetation terms are the same as Shinyiha. Malila agricultural systems are also similar to the Nyiha. *Nkule* fields are made both in upland grass and *mbuga*. The Safwa-type of *ishizi* with grass-burning is called *isengo* among the Malila; *insavi* is also made. The Malila once made numerous *insenjela (nkomanjila)* which are now outlawed.

The Lambya made *ntemele (amatemwa)* in the past but the dominant contemporary field type is *insimpa* which is essentially the Nyiha *itindiga*. *Nkule* is not made, and *nkomanjila*, while known, is only rarely used. Second-year fields are known as *lisala*, and subsequent crops will often be ridges *(mandi)*. Among the Nyakyusa the dominant field form is the large contoured ridge whose drainage and erosion-prevention features are clearly related to the heavy rainfall of Rungwe area, totaling up to 120 inches per year. The major staples are plantains and ridge-cultivated maize and sweet potatoes. In his description of Nyakyusa agriculture, Thwaites (1944) described an unnamed slash and burn method that was practiced in the hill land of western Rungwe perhaps among the Ndali. This is probably the *nkomanjila* system. In addition, the Nyakyusa grow finger millet on grassland fields prepared by burning and hoeing.

An additional cropping system not previously discussed is the Matengo pit system of hoeing (Pike 1938). The Matengo live to the east of Lake Nyasa (Figure 2.1), concentrated in highland areas by Yao, Ngoni, and Arab slaving raids. There, according to Pike, they were faced with the problem of producing food on an indefinite basis from limited land resources. They evolved an ideal system for coping with steep hill slopes with virtually no erosion. Grass is cut close to the ground and laid in perpendicular 2-foot strips about 6 feet apart forming a rectangular lattice. Earth from between the strips is hoed onto the grass, creating a network of square, closed pits. Crops are planted on the mounds, and weeds and refuse thrown into the pit. In the subsequent year, the whole lattice of mounds is shifted so that refuse is incorporated into a new mound lattice. Runoff accumulates in the pit and slowly soaks in. Pike noted that even the smallest streams do not fail during the dry season, probably as a consequence of the decreased water loss to direct runoff (Pike 1938). I did not see any evidence of traditional use of this system in Mbozi or neighboring areas. However, the government has made extension officers aware of the system. I found one young coffee field in Nyimbili VDC in which the field had been hoed into pits in which the trees were planted. The farmer had been told to do this on the steep hill by *bwana shamba*. The field was too young for him to judge whether his coffee yielded better than his neighbors'.

Several basic agricultural traditions in the Corridor Area are widespread (Table 6.1). These are the grassland mounding systems (Trapnell's Millet Grass Culture), *citemene* varieties, the *nkomanjila* system, and the ridging systems. The basic *nkule–intumba* and *citemene* systems are probably of greatest antiquity. The *nkule* method may have evolved from the *intumba* system as an adaptation to the more acid soils of the *mbuga* margin.

Table 6.1

Distribution and Evolution of Agricultural Systems–Tanzanian "Corridor" Peoples[a]

	Grassland Mounding Systems (Millet Grass Cultures)		Woodland Systems (Millet Ash Cultures)	
	Intumba	*Nkule*	*Nkomanjila*	*Citemene*
Fipa	X			
Mambwe	X			X
Namwanga	X	X		X
Wanda	X	X	X	X
Nyiha	(X)[b]	X	X	X
Lambya			X	X
Safwa		X	X	
Songwe		X	X	
Malila		X	X	
Nyakyusa			(X)[c]	

[a]The use of each technique is indicated by society.

[b]The Nyiha used *intumba* in the past.

[c]The Nyakyusa have a number of other agricultural systems in addition to *nkomanjila*.

There, even after burning, the soils are sufficiently fertile that a rapid nitrogen flush can support immediate cultivation of grains without a leguminous catch crop. As applied to upland grasses, *nkule* seems to be limited to the volcanically derived soils of Mbozi, Mbeya, and Rungwe areas. In these areas of higher rainfall burning is an important weed control measure. *Nkomanjila* may have evolved from the *citemene* system in southwestern Tanzania in the presence of fertile soils. Here, more luxuriant crop growth militates against weeds even if the cut trees are not hauled for burning to a limited area of ash concentration. *Ntemele (citemene)* may have had a resurgence when only secondary *miombo* bush was available to be cut and lower crop yields increased weed problems. Ridges (*matuta, mandi;* not shown in Table 6.1) are very widespread; it is difficult to suggest their origin. Where they are not used in major field sequences, ridges are used in subsidiary fields or gardens. In gardens they were probably an early response to soil drainage problems and may represent a morphological evolution of earlier mounding systems.

Culture Contact and the Spread of Ideas

In an attempt to confirm which Nyiha agricultural practices were traditional as claimed rather than recent arrivals, Nyiha informants were asked how neighboring peoples make fields. I felt that if they were able to identify accurately the neighboring systems, then further analysis of whether the analogous system in Unyiha was a recent arrival would be warranted. However, the Nyiha were generally vague and inaccurate about their neighbors' agricultural practices unless there happened to be a person present from a particular society or a Nyiha who had visited there. Informants at Wanishe were not aware of the Namwanga making *ntemele*. At Nambinzo an argument among informants ensued over how the Safwa prepare fields. While eastern Nyiha were aware of nearly identical Malila, Safwa, and Lambya agricultural techniques, in Nambinzo one group of informants knew nothing about agriculture among these peoples. In Msia I was told that the Fipa make nothing but *nkomanjila*. Thus there was no consistent pattern from which recent arrival of the major traditional systems could be inferred. There was one consistency, however, in attributing large ridges to the Nyakyusa and Ndali, who were always credited with introducing this practice to Unyiha. Thus most agricultural similarities we see today between the Nyiha and their neighbors reflect their common past or diffusion at an early period. Hence, it is valid to infer that the traditional Nyiha *nkomanjila* and *nkule* methods are indigenous from the point of view of living Nyiha.

Migration and Intercultural Marriage Patterns. The method by which the Nyakyusa ridging practices were brought to Unyiha has already been mentioned—migration in search of land and jobs. While Nyiha speak of some introductions of crops by those returning from labor migration (the mango, for example), there was no indication of farming practices having been recently brought by this method. However, this does not rule out the possibility of Nyiha traders having brought agricultural techniques during the precolonial period. Some intercultural marriages have taken place in Mbozi. Informants suggested that marriage with the Namwanga, Malila, and Lambya is common. That women from another society could bring considerable agricultural knowledge is obvious. Whether the Nyiha neighbors had much new to offer in the living generations is doubtful. In recent years, at least, intercultural marriage has not been agriculturally significant. I met few men with other than Nyiha wives.

Language Barriers. The language barriers discussed in Chapter 2 would have an obvious effect on the interchange of agricultural lore among peoples. However, the common core of agricultural knowledge seems to have been established before the present linguistic barriers evolved. Today Kiswahili crosses these barriers, but its major role is in the dissemination of officially sanctioned agricultural innovations rather than spontaneous introduction and diffusion of crops and techniques.

Population Pressure and Land-Use Intensification

Ecologically, it is reasonable to argue that the European Intervention in Mbozi with accelerated population growth and cash crops did not alter the ultimate course of traditional agricultural change but simply hastened it. Land allocated to the estates was removed from the supply of land available for the traditional cropping systems. With regard to food production, allocation of nonestate land to cash crops is analogous to population growth. Over a longer time period, the evolution of the Mbozi Plateau *Parinari* cultivation steppe would still have occurred. The resultant emphasis on grassland agricultural systems would still have been the response to the need for intensification. Hence, the coming of the European has not altered a basic viewpoint toward agricultural change, but in fact compacted the long-term process and made it more readily scrutinized. Here, in Mbozi we have a model of what would have resulted had the growing traditional society remained isolated from the Western world. However, intensification is only one aspect of agricultural change in Mbozi. Alien sources have brought change of other dimensions.

The Alien Sources

The colonization of Tanganyika by the European powers—first the Germans and subsequently the British—made possible the contact that brought the Nyiha into full interaction with the wider world. Direct military activity, alliances with African peoples, and cooperation from missionaries enabled the Germans to subjugate the population of the northern Corridor region. The development of German rule, and after World War I, British rule, set a stage for accelerated change. This change is reflected in social and political activity and, for our discussion in particular, agricultural patterns. All over Unyiha a wide variety of crops (European potato, guava, papaya, citrus, onion, cucumber, lettuce, tomato) are attributed to one man, the founder of Mbozi Mission. Settlement of Mbozi by European planters was a direct result of government policy and led to the diffusion of coffee cultivation as the major cash crop. The Asian merchant community facilitated African cash crop production by providing the initial market outlet. Although this role was later taken over by the cooperatives, the trader and his store remain the focus of economic desires and market dependency. Here, we outline some of the general background and local consequences of the alien intercession, moving from the German period and the establishment of missions to the British succession, creation of the estate economy, and development of the merchant community.

The German Period

German interest in East Africa was manifested in the middle-19th century trading houses in Zanzibar and early penetration of the interior by missionaries and explorers. Establishment of German East Africa has been summarized by Freeman-Grenville (1963:434–435):

> Between 1885 and 1887 little more was done than to establish a series of fictitious treaties with allegedly sovereign chiefs. When, between 1888 and 1890, the Germans began to extend their influence over the coastal region, they were faced by an uncoordinated by desperate resistance [near the coast]. . . . It was not until after this rising had been subdued that any serious attention could be paid to the effective occupation of the interior, and this was concluded in 1898 with the death of . . . the chief of the Hehe.

Many of the general policies of the Germans were felt in Mbozi and vicinity. Arrangements were made with missionaries to represent the government, and alliances with African peoples were sought against others yet to be subjugated. From 19th-century trade networks the Germans built a communication network that indelibly wrote patterns of spatial interaction on the Tanganyika landscape. Kiswahili was selected as the territorial language. Permitting continued mission activity, Germans also strengthened secular education. As qualified men emerged, they were brought into government service, a valuable return for the flow of African labor required to support colonial economic activity (Wright 1968b).

The first German settlement in southwestern Tanzania was constructed at Langenburg, a Nyasa shore site selected as most accessible to the dense population concentration at the northern end of Lake Nyasa. The name of the settlement was derived from Prince Hohenlohe-Langenburg, a promoter of German colonial endeavors. The lake bottom drops sharply along the northeastern coast of the lake, so that excellent moorings could be obtained. A fort *(boma)* was built of logs, followed by permanent brick houses and gardens (Wenban-Smith 1963). Protestant Moravian missionaries had arrived in 1891 and established themselves at Rungwe. In 1899, the German mission at Mbozi was founded and a small fort was constructed at Itaka on the route from Langenburg to Bismarckburg in 1900. The district office of Langenburg was moved to Neu-Langenburg (now Tukuyu) in 1900. The Germans rebuilt the old Stevenson Road which the British had constructed linking Lakes Nyasa and Tanganyika. Cultivation of coffee, rubber, and other nonlocal crops was begun at Rungwe, Mbozi, and other Moravian Missions, as well as by a small group of planters in three areas: the present Rungwe Area, the present vicinity of Mbeya, and one farm in Mbozi near the Ruanda River (Great Britian Admiralty 1916; Letcher 1918; Nowack 1937; Gemuseus 1938; Raum 1965).

The Mission

In 1891 German Lutherans and Moravians had entered the East African mission field together, dividing the region along the 34th meridian. The Moravians were to proselytize to the northwest of Lake Nyasa and the Lutherans to the northeast. By 1914 there were a total of 30 Moravian missionaries in 15 stations among the Ngonde, Ndali, Safwa, and Nyiha, including the Mbozi mission established by Traugott Bachmann. Like other missionaries, Moravians believed in economic as well as spiritual rebirth.[3] Africans were employed to help build the missions and to work in the fields. Bachmann took workers from each of the 12 Nyiha chiefdoms in rotation for a month to work at the mission. In this way he introduced many to work and worship at the mission without offending any chief (Wright 1971:105). At the same time new crops were carried rapidly to the Nyiha, perhaps an unintentional by-product of Bachmann's strategy. Native evangelists were trained to carry the message beyond the mission station. In 1916 Bachmann was expelled from Mbozi by the British (Bachmann 1943). Large-scale farming on the mission ended with the departure of Bachmann. From 1916 to 1925 the Moravian missions were run by the Scottish Presbyterian Livingstonia Mission from Nyasaland. In 1925 the Tanganyika government allowed the return of ex–enemy missionaries and restored their former property. Except during World War II, Mbozi was served by a number of German Moravians from 1926 until 1959. African clergy served the mission during World War II, and it was completely Africanized in 1959 (Bachmann 1943; Moffett 1958; Mwalupembe 1967).

Many missionaries were given the names of local chiefs. Bachmann was named *Mwalwizi* while in Rungwe District and became widely known by that name. Of the 300 acres he purchased from Nyiha chiefs Mwasenga and Nzowa, Bachmann planted about 20 acres to coffee for export. Some of these trees are still growing on the mission grounds. Bachmann experimented with many other crops, introducing them to the Nyiha coincidentally. The way in which Africans learned from Bachmann is analogous to the process to take place later on the estates. Africans hired by Bachmann were trained to cultivate new crops and use new instruments like the plow. These ideas were taken home and, when successfully instituted, spread to other Nyiha.

Mwalwizi's success was due in part to the open mind with

[3]The role of the missionary in East Africa has been reviewed by Oliver (1966), Johnson (1967) and Wright (1971).

which he learned Shinyiha and Nyiha customs. In his proselytizing, Bachmann moved from traditional Nyiha belief toward Christianity, rather than attempting wholesale substitution (Wright 1971:105–107). His biography provides an excellent review of Nyiha cultural tradition at the turn of the century, and indicates his great sympathy for customs that seemed superficially strange. For example, he learned that the bride-price did not represent "sale" of women by their parents. Bachmann found that the availability of cash labor attracted more Nyiha than he could employ, even though in 1 month he employed 300. He was impressed by Nyiha diligence and perseverance for even low wages (Bachmann 1943:124–125, 139–141).

In addition to new crops and agricultural techniques, the mission left the indelible imprint of Christianity and education. Christianity may have been a strong catalytic factor in the early acceptance of coffee. Formal education was established, and numerous villages were visited by evangelists forming bush schools or catechetical centers, preparing the groundwork for the contemporary educational system and the role it plays in further fostering change.

The British

The British captured the southwestern part of Tanganyika during World War I by invasion from the south. The civilian staff had retreated with the German army, so British political officers manned the German *bomas* (Hatchell 1958). Southern Highlands Province was created during the war, and in 1919 Rungwe District was organized with most of present Mbeya and Mbozi Areas. After the war, German East Africa was given to the British under a League of Nations mandate (1920) and the separate southwestern administration was absorbed by Dar es Salaam. Mbeya was established as a district office with a district officer in 1926 (Tanzania IPB). In the same year, tribal boundaries were fixed throughout the province and tribal governing authorities established.

In response to the necessity of maintaining the estates and other foundations created by the Germans, the colonial government yielded to pressure for continued European settlement in Tanganyika (Ingham 1965:548–550). The mandate from the

League of Nations did not preclude such settlement, requiring
however explicit assent from the native peoples (Harlow
and Chilver 1965:692):

> No native land may be transferred, except between natives, without
> the previous consent of the public authorities, and no real rights over
> native land in favour of non-natives may be created except with the
> same consent.

The Land Ordinance of 1923 provided for leaseholds rather than
freeholds on land, and in 1925 a ban on the return of German
nationals was lifted. A number of these, plus settlers from Britain,
other European areas, and Kenya arrived to take up land. As
we shall see, Mbozi was a critical locus for this colonization.

The administration of Sir Donald Cameron (1925–1931; Ingham
1965) was important to all of Tanganyika for its articulation of
indirect rule, its attitude toward European settlement, and its
affirmation of African production of cash crops. Indirect rule,
as we saw with regard to consolidation of peoples, was in some
ways more ideal than real, yet the intention to build, in general,
from existing institutions was realized. Although Cameron continued to assert the primacy of African interests, he recognized
the potential benefit settlers might have in the development
process, and incorporated them within the Legislative Council
system. The Mbozi alienations were thus completed in a political
milieu hospitable to European settlement, creating in Mbozi an
indirect administrative impetus for development. Finally, African rights to grow cash crops, as we have seen, were asserted
under Cameron.

In 1922 an English trader who heard rumors that Germans
had earlier found gold in the hilly area east of Lake Rukwa
discovered alluvial gold himself on the Lupa River. By 1925
97 claims had been staked in the Lupa area, and in 1932 the
Lupa Controlled Area was created to administer the diggings.
Most of the output was from reef production by that time, and
reached a peak of almost £300,000 (then $900,000) in value in
1941. By 1948 the area was in a depression; the last mine closed
in 1956. Some gold can still be found in Lupa. Africans will
occasionally offer the precious commodity for sale.

Lupa attracted not only European gold diggers but also African
laborers, many of whom were not local but from Rhodesia,

Nyasaland, or the Bena or Nyakyusa tribes (Tanganyika ARPC 1937:64). Lupa and estates provided a market to which Asian merchants responded by establishing stores. One of the present Mbozi planters is a descendant of a Greek who supplied hardware for farm and mining trade. Lupa also meant a small but steady market for food; local peoples, including the Nyiha, were induced to sell surplus commodities to Asian merchants who marketed in Lupa and elsewhere. Hence, while Lupa today is an historical curiosity, it played a brief but important role in the beginnings of rural change and development.

The legacy of the British can be seen in the emerging political system and the fostering of settler–African interaction leading to development, along with continued improvements in transportation, communication, administration, and health services.

The Mbozi Planter

The successful establishment of the mission at Mbozi had indicated that the plateau area was suitable for European occupation. In 1906 Fülleborn noted that the area had great potential for settlement since a number of crops did well and the wide *mbugas* could be used for cattle production (Fülleborn 1906:477). Shortly after World War I, alienation of the Kenya Highlands was completed, and farm prices there climbed. Lord Delamere, a proponent of white settlement in Kenya, had turned his attention to Tanganyika and encouraged settlement in the Iringa and Sao Hill areas. He also noted the potential of Mbozi. Hence, many potential settlers disappointed by the lack of land in Kenya turned south (Jacobsen 1951, 1954; Vernon 1968). The opening of Tanganyika for settlement and the post-1925 return of Germans was well-publicized in Western Europe and Africa.

Land made available in Mbozi by agreement with Nyiha chiefs was rapidly alienated from 1925 until Government Notice 1135 on 22 December 1927 closed all of Mbeya District to further alienation (Great Britain Colonial Office 1939). This period was sufficiently long for many settlers who were intent on Mbozi to take up land, along with several "accidental" arrivals. Alienations ranged from 190 to 2000 acres. Some farms were later subdivided, and in 1939 there were 63 estates in Mbozi.

While a few of the settlers were sufficiently wealthy to sustain the 4 or 5 years until coffee came into production, others set about mixed farming—raising pigs, cattle, turkeys, and chickens in addition to arable crops and coffee (Jacobsen 1951, 1954). Experts who visited Mbozi were less enthusiastic about coffee production than the settlers. Gillmann (1927) had doubted the suitability of Mbozi for coffee due to the dry season, and in a later report (1929) sounded a dealth knell of any extension of the Tanganyika Railway to the southwest that was to prevail for 40 years. In a 1928 article, the Acting Director of Agriculture included a guarded assessment of Mbozi's potential (Wolfe 1928:727–728):

> Mbosi. Altitude about 5,350 ft. The rainfall is judged to be about thirty-eight inches. The principal soils are (a) a rich clay loam, with a very high capacity for moisture retention; (b) a very light brown sandy loam. The former is suitable for perennial crops, the latter annual . . . The settlers intend to plant coffee. On the red clay loam this crop would have quite a good chance of succeeding, but carefulness is advisable as . . . dry atmosphere and low temperatures in the winter, as well as excessive variation between day and night temperatures in the winter, may have an injurious effect, though these factors are not so unfavourable at Mbosi as at Mbeya. Heavy shade will be necessary . . . . Extensive plantings initially are not advisable.

In a report on Mbozi agriculture prepared by A. E. Haarer (an expert on coffee production and, at the time, District Agricultural Officer at Moshi), Mbozi was stated to be quite unfavorable for coffee due to its "uneven rainfall" and extreme temperatures. Haarer felt heavy shade, irrigation, and manures were absolutely essential. Coffee in Mbozi, according to Haarer, was totally experimental (Haarer 1928).

By 1932 Mbozi coffee was beginning to appear successful, while many other areas of the Southern Highlands were giving it up. Mbozi coffee was established as the highest quality in Tanganyika in 1934, an approbation that remains valid today (Tanganyika Department of Agriculture 1933, 1935). Mbozi estate coffee commands a premium price and is sought mainly by German buyers for blending purposes. The inaccuracy of some of the early predictions about Mbozi was probably due to an underestimation of the rainfall.

Over half of the Mbozi settlers were Germans. Some attempted mixed farming, while others invested all their capital in coffee plantations without sufficient trials or knowledge of coffee management. By the outbreak of World War II, many of them were severely in debt to the Reich-financed Usagara Trading Company and its subsidiary, the Uhehe Trading Company. Many Germans on contract to the Uhehe concern were given a cash allowance of 400 shillings per month (then about 60 dollars) and surpluses due from coffee delivered to the company were paid in goods. Moffett (1958: 119) remarked:

> In fact, one of the few successful German coffee farmers in the Mbozi area was well known as having the finest set of farm buildings in Tanganyika while he himself lived miserably on the . . . cash allowance . . . having been supplied with cement and corrugated iron far in excess of his needs.

At the time of property confiscation in 1939, Germans held 90% of the hotels and garages in the Southern Highlands Province.

German farms (Figure 6.7) confiscated during World War II were managed by the Custodian of Enemy Properties. After

Figure 6.7. A prewar German farmhouse. Many of the German settlers were poor. Some of their homes were reoccupied by new settlers and eventually improved or replaced after the war.

the war a few were returned to native authorities, while the remainder were realienated to soldiers and civil servants. Some of the farms were consolidated to become economically viable, and others called "rentier" farms were intended to supplement a reliable retirement income.

In the early 1950s, the planters hoped to attract a National Development Corporation ranching scheme to Mbozi. This was rejected due to the large amount of African settlement, distance from the rail line, and the limited availability of land (Tanzania Archives File 40511/11). The settlers then proposed establishment of five new estates in the Ruanda–Mahenje area. This was rejected by the Governor in 1954 at least partially in response to Native Authority resistance (Tanzania MbDB). Still, the settlers, through the Mbozi Farmers' Association and the Mbozi Club, a social organization, had been able to secure a post office, dispensary, agricultural experiment station (Mbimba), subdistrict administrative offices, and police post through their articulation of a comprehensive Mbozi Development Plan. While the altruism of the settlers' activities could conceivably be questioned, their accomplishments as a powerful pressure group benefited the Mbozi community as a whole.

Yields on Mbozi estates have never been spectacular. Two hundred pounds of parchment coffee per acre has been considered a good yield. By the 1960s, many of the postwar settlers had suffered financial defeat. Only those able to afford large capital inputs of chemical fertilizers and supplemental irrigation have done well. Many of the estates were sold to Asians; some were even abandoned (Figure 6.8) in the face of insurmountable debt. Hence, in the late 1960s, economically viable yields of 500–1500 pounds per acre are found only on the farms with the new irrigation and fertilizer technology evolved by the Mbimba experiment station and enterprising planters.

The outlook for the European planter in Mbozi is poor. Many of the farmers suffered from a paranoia over the political situation. Only one or two have been willing and able to make sufficient investments to make their estates truly profitable. Others have been anxious to sell at virtually any price, while a few are enjoying their remaining years in a pleasant existence. The whole situation has been exacerbated by falling

Figure 6.8. An abandoned European farmstead. Within 1 year after abandonment the dwelling had been completely disassembled for brick and timber, leaving only the chimneys and decorative plants behind. The coffee plantings will also be lost as bush fires destroy the trees.

coffee prices and fear that acreage restrictions might be imposed. If the era of the European estate is near an end, its impact on the Nyiha continues.

Farm Dispersion. One of the most important features of land alienation in Mbozi was that no single block of land was set aside for exclusive European use. Rather, consent of local native authorities was obtained before alienation was allowed to begin. They agreed on the condition that no further alienations would be requested (Bagshawe 1930:14). Once prospective settlers chose a site, each choice was subject to tribal approval as well as that of any African living there. The latter were compensated on a fixed schedule for their improvements. They were reimbursed for houses and grainstores constructed and fields in cultivation that would be displaced by the estate (Tanzania Archives File 33//L2/45).

From the settler's point of view, dispersion was a mixed blessing. It assured that a number of potential employees would be located near the farm. As a consequence, housing and food did not have to be supplied. On the other hand, the European community was dispersed, and gatherings on social or business occasions were difficult using the barely passable roads during the wet season. Pilferage from the plantation was a menace; even today unarmed guards are hired to patrol the farms as the coffee ripens for picking. This problem increased in magnitude after African coffee production became more common. Indeed, the inability of authorities to identify the source of coffee sold supported the planter's efforts to have African cultivation of coffee declared illegal.

Farm and Labor Organization. The Mbozi planter has been dependent upon large inputs of relatively cheap labor in order for his estate to remain profitable. The typical coffee estate employs one permanent laborer for every 2 or 3 acres of coffee in production, plus cattle herders, tractor drivers, and household help. During the picking season, some two or three people are employed on a daily basis per acre. Many of the permanent laborers work full-time only in the picking season and half-time during the remainder of the year. Pre-World War II wages of 10 shillings per month had grown in 1967 to 80 shillings per month for full-time laborers. Given the relatively low yields and high transportation costs, the estates would probably have met financial crisis sooner had wages been higher.

The Ecology of Cash-Crop Change. As in the case of the mission, introduction of coffee and techniques for its cultivation to the Nyiha is centered on the estate employee. The estate provides training and at least some could afford experimentation. Results are carried to the personal plantations of estate laborers and spread from there to other coffee cultivators. While the scale may be altered, the same basic techniques are applied. For example, spraying of copper sulfate and insecticides on the estates is usually accomplished using tractor-borne power sprayers. The same technique is emulated using inexpensive hand-operated knapsack sprayers on the small farm. Many Nyiha informants attributed their knowledge of pruning and mulching to neighbors who were working or had worked on estates.

This process of change is only possible because the Nyiha can take what they have learned to an analogous environment. In Mbozi, the seasonal chores on the estate could be repeated on the home farm. Hence, the dispersal and day-to-day interaction of estate and small holding through cash laboring comprises a viable ecology of voluntary change in cash-crop technology. The historical pattern of coffee production reflects this process (Figure 6.9). In 1945 only near the estate areas was coffee grown in at least 10% of the *Kumi-kumi* groups which now constitute a village development committee.[4] In 1955, these early Nyiha coffee areas indicated greater intensity of coffee cultivation than elsewhere, a pattern that finally disappeared in 1967 by which time coffee was ubiquitous. Still, only one-half of all Nyiha families were growing coffee in 1967, so that continued expansion of coffee production is likely.

The important contribution of the spatial juxtaposition of estate and the Nyiha to rural development can be illustrated by comparison with the former "White Highlands" of Kenya (Huxley 1953; Morgan 1963; Bennett 1965). Lack of a restrictive mandate and European pressure led to creation of the block of land from which African settlement was excluded. Laborers came from some distance to the highland farms, stayed there for long periods, and were provided with foodstuffs and housing. Not only was transfer to the home area of agricultural techniques learned ecologically improbable, European political pressure succeeded in formulation of restrictions against African production of certain cash crops. Thus the African laborer was very unlikely to carry substantial agricultural information home with him. Moreover, he was prevented from growing many of the crops with which he had worked. Thus, the spatial and political organization in Kenya

---

[4]This contemporary perspective does not, of course, tell us about the circumstances within which the very first people engaged in cash-crop production and initiated the modernization process. Brock (personal communication) has suggested that these initial innovators may have been Christians and were of high social status. My limited information on this point confirms her observation in reference to some of the earliest coffee-growing areas, Iyula and Nsala for example, whereas in the areas I visited that had begun coffee cultivation somewhat later, there was a direct linkage from estate employee to his smallholding and to his neighbors.

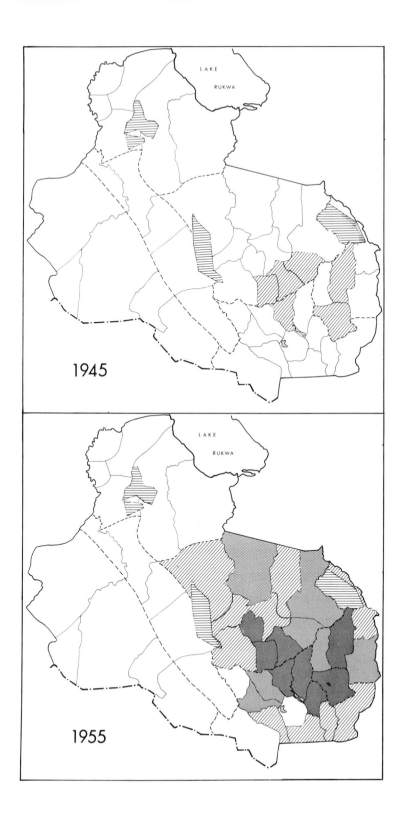

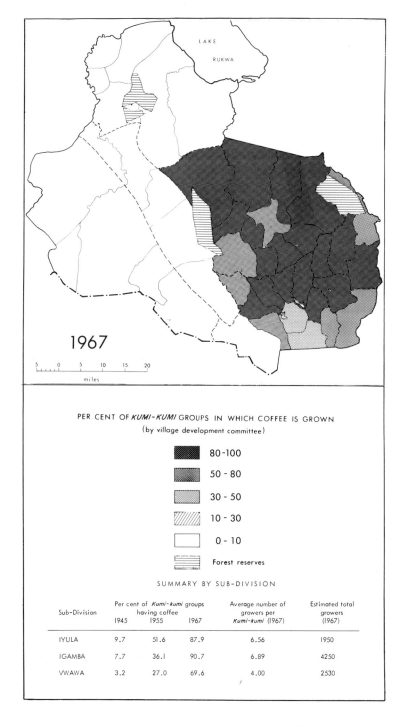

Figure. 6.9. Coffee in Mbozi Area—1945, 1955, 1967. [Data from Mbozi Economic Survey, Appendix 2.]

may have effectively precluded rapid diffusion of crops and technology to African peoples, while, in contrast, the spatial arrangement of estate and African in Mbozi would appear to have been designed to implement such a mechanism of change!

The Asian Trader

Active as businessmen and financiers on the East African coast under the Arabs, Indians were also employed by the Germans as artisans, mechanics, and civil servants. The British actively recruited Indians for the civil service after World War I while others came of their own volition in various entrepreneurial roles. A large portion of them became merchants, and by establishing stores (*dukas*) in rural areas, were often the first agents of the cash economy (Moffett 1958:298–302; Livingstone 1965).

The first Asians arrived in Mbozi in the 1930s. A small trading settlement grew up at Old Vwawa where the Great North Road makes its first river crossing southwest of the present Vwawa. By 1943 there were five Asian families operating small stores and acting as agents for purchasing African produce and providing trucking service to Mbeya. In 1942 an official trading center was established at Tabora near the Mbozi Mission. The Old Vwawa center rapidly died in favor of Tabora. In 1954 a new Vwawa settlement was planned on the realigned Great North Road. The first building began in 1955, and the official government substation opened the next year. The Asian community began moving to Vwawa, the last merchant leaving Tabora in 1959. An additional trading center has grown at the international boundary crossing at Tunduma.

The East African Asian community as a whole is not homogeneous but consists of peoples of varying ethnic and religious origins (Delf 1963; Ghai 1965). The Mbozi Asians, however, are predominantly of the Ismaili sect of Islam. A number of the local families are related and most have relatives in Mbeya. Many were born in Tanzania and have taken local citizenship since independence. There is a cameraderie of support and competition that links the business community together. In addition, they are tied by worship in the Jamat Khana in Vwawa. Their children were educated in the Aga Khan school which now serves African as well as Asian students and is government supervised. Virtually every member of the Asian community stated that business was better for all by being in one community rather than scattered over the countryside. However, several families did live in Itaka and Ruanda as well as Tunduma and Vwawa. The Asian community is no longer limited to business roles. Several families, one of them from Mbeya, have purchased coffee estates. Most have been moderately successful in these ventures.

The role of the Asian trader in Mbozi has been to buy African crops and at the same time to purvey goods for sale, partially satisfying needs he himself helped to create. He may sell directly or finance African *dukas* and sell goods to them for resale.

Sale of Commodity Surpluses. Until the establishment of cooperatives, the Asian community provided the means by which Nyiha farmers could sell produce and indirectly reach larger markets. The major crops marketed through the Asians were surplus quantities of traditional foodstuffs (finger millet, maize, beans) and subsidiary crops (sesame, castor). Most of the produce went to Mbeya, some to the Lupa gold fields, and some as far as the Zambian Copper Belt. Although the total quantities of produce handled by the Asians cannot be established, one man stated that he had purchased 200 tons of sesame, 120 tons of castor seed, 100 tons of maize, 100 tons of beans, and 200 tons of finger millet in 1957. Another said he had sent 100 tons of beans to the Copper Belt that year. If these recollections are reasonable, the Asian community made a significant contribution to cash-crop production in Mbozi.[5]

Rising Desires and Expectations. In exchange for crops or money, the Asian offered the wares of the world: cloth from America, England, Holland, India, Japan; metal tools from England and the Continent; tin lamps from India; enameled East African dishes; soap; salt; sugar; Coca-Cola; cooking and lighting oil; thread; matches. He usually offered the services of a tailor and was willing to partake in the palaver of bargaining that made shopping a social event. As the rural com-

[5]Early mobilization of traditional crops may represent disposal of the "normal surplus" that was frequently harvested during good years, but would represent plantings sufficient to provide assurance of satisfactory food supplies during poor years (Allan 1965:38–48).

munity became more wealthy in the 1950s, he became the main supplier for African bush *dukas* (Figure 6.10), although he often sold to them at retail rather than wholesale. The bush *duka* has carried the newly desirable goods even closer to the people. The trader and the planter would be emulated as models for change. They became the focus of a circular growth process by providing the means and opportunity to buy. The bicycle and radio (Figure 6.11) are part of this ecology. Both provide increased accessibility to new ideas. Hence, they contribute to spiraling dependence on the *duka*.

That today's luxuries become tomorrow's necessities hardly needs to be stated. This certainly holds in Mbozi where a large variety of household and personal items once manufactured in the home or less commonly by village specialists are now purchased in the *duka*. The iron hoe is a typical example. Local smelting of iron has ceased and will, in all probability, become a lost art (Brock 1966). Hoe blades are purchased, costing from 6 to 12 shillings. Virtually no clothing items are homemade; sugar and salt are bought in the *duka*; and for many families, meats other than chicken always come from the butcher. Articles such as clothing, bicycles, kerosene stoves, and lanterns are no longer simply items of conspicu-

Figure 6.10. A bush *duka* or general store. Although established with very little capital, the *duka* becomes more adequately provisioned as local prosperity increases.

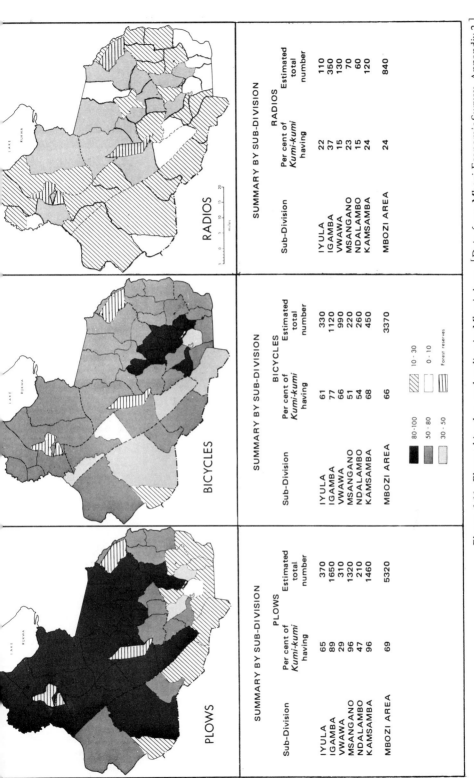

PLOWS

SUMMARY BY SUB-DIVISION

| | PLOWS | |
Sub-Division	Per cent of *Kumi-kumi* having	Estimated total number
IYULA	65	370
IGAMBA	89	1650
VWAWA	29	310
MSANGANO	96	1320
NDALAMBO	47	210
KAMSAMBA	96	1460
MBOZI AREA	69	5320

BICYCLES

SUMMARY BY SUB-DIVISION

| | BICYCLES | |
Sub-Division	Per cent of *Kumi-kumi* having	Estimated total number
IYULA	61	330
IGAMBA	77	1120
VWAWA	66	990
MSANGANO	51	220
NDALAMBO	54	260
KAMSAMBA	68	450
MBOZI AREA	66	3370

80-100   10 - 30

50 - 80   0 - 10

30 - 50   Forest reserves

RADIOS

SUMMARY BY SUB-DIVISION

| | RADIOS | |
Sub-Division	Per cent of *Kumi-kumi* having	Estimated total number
IYULA	22	110
IGAMBA	37	350
VWAWA	15	130
MSANGANO	23	70
NDALAMBO	15	60
KAMSAMBA	24	120
MBOZI AREA	24	840

Figure 6.11. Plows, bicycles, and radios in Mbozi Area. [Data from Mbozi Economic Survey, Appendix 2.]

ous consumption. As de Wilde (1967, 1:59) has suggested, this regular standard of consumption is likely to assure continued economic motivation for increased production.

Today the cash economy extends beyond the Asian *duka*. In order to ascertain the extent of development in this respect, a survey of all economic establishments in Mbozi was conducted in 1967 (Appendix 2). Categories of businesses in Mbozi include *dukas*, butchers, power grain mills *(posho* mills), restaurants, and bars. The distribution of these businesses reflects both the concentration of wealth in Unyiha and centralization at settlements and governmental centers (Figure 6.12). The general living standard of a Nyiha family selling crops or labor is not high (Table 6.2). For a typical Nyiha family a bicycle represents more than a year's cash income. Income must be reduced by the "local

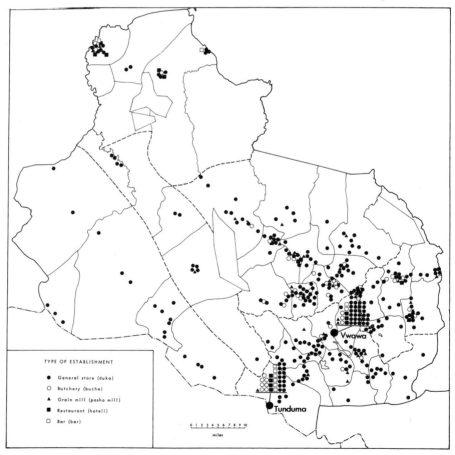

Figure 6.12. Business establishments in Mbozi Area, 1967. [Data from Mbozi Economic Survey, 1967.]

Table 6.2

Mbozi Price Structure—1967 [a]

I. Income

Full-time farm labor per month	80
Household help per month	100
School teachers per month	300–700

II. Crops and Livestock

Cattle	100–200	Finger Millet *(debe)*	5 (July)–12 (March)
Sheep	25–40	Maize *(debe)*	4(June)–10 (Jan.)
Goats	25–40	Beans *(debe)*	10
Donkey	100–150	Wheat *(debe)*	10–12
Hen	1–4	Coffee (pound)	1
Rooster	3–6		
Pigeon	1–2		

III. Local Products and Services

Ibande house with thatched roof	100–150	Ax handle	1
100 sun-dried bricks	1–2	Hoe handle	2
100 baked bricks	3–4	Reed mat	5–10
Carved stool	1–2	Basket	1–3
		Forged ax blade	2

IV. Imported and Nonlocal Products

Corrugated metal for roof, sheet	6–9	Kanga (women's cloth wrap	15–20
Blanket	15+	Lady's dress	25–40
Paper, 100 sheets	5	Child's dress	10–12
Sugar, pound	1	Men's shorts	10–15
Wheat flour, 5 pounds	3–4	Men's shirt	10+
Rice, 5 pounds	3	Plastic shoes	10+
Hoe blade, import	6–12	Leather shoes	30+
Meat, pound	1–2		
English bicycle	350–450		
Plow, single share	100		

[a] Prices shown were observed in Mbozi in 1967 and represent final prices after bargaining. Values are expressed in Tanzanian shillings, each worth 14 US cents or about 7 shillings to the dollar.

rate" tax, a poll tax amounting to 45 shillings or 1 month's labor for the District Council. Even if one assumes that all the cash crops sold through the Mbozi cooperatives originated in Unyiha, the average family income from this source would only be about 18 dollars per year. Although this income is supplemented by contributions from estate labor or labor migration, by any measure of monetary wealth, the Nyiha are, on the whole, very poor.

The Government

Both the colonial and independent Tanzanian governments have been agents of agricultural change. These changes have resulted directly and indirectly from government policy. Indirectly, imposition of taxation forced the Nyiha into the labor market or cash-crop production. Through laws aimed at protection of natural resources, through regulation of forest cutting, and through agricultural extension activities, official policy has directly affected agricultural practices and change.

Indirect Change: Taxation

In 1897 the Germans imposed a hut tax in East Africa, followed by an annual poll tax in 1908 (Henderson 1965:133). The result of these taxes and their continuation under the British regime was to force the Nyiha (and others) to produce crops for sale or to enter the labor market. The present tax structure assures continuation of this impetus.

The major tax faced in 1967 by all Nyiha men was a uniform, annual 45-shilling local rate imposed by the District Council. This could be paid in cash or by labor. Those avoiding both means were subject to imprisonment. The national Tanzanian personal tax applies only to those with an annual income over 2000 shillings per year and is graduated from 1 to 3.75%. The national income tax which is graduated from 10 to 75% applies to single men with incomes over 4500 shillings and married men whose income exceeds 14,000 shillings (Lee 1965). There are a number of Nyiha liable to both personal and income taxes, but the major incentive to economic growth has been the indiscriminately applied local rate which forced Nyiha men into the market economy.

Direct Change: Agricultural Regulations

Among the earliest government activities in the traditional agriculture sphere was a series of prohibitions aimed at the government's interpretation of the destructiveness of traditional methods. The following laws promulgated in Mbeya District in 1938 (Tanzania Archives File 77//2/33) illustrate these provisions:

1. It is an offense to allow a fire to escape from a field
2. It is an offense to set fire to vegetation outside a field
3. It is an offense to allow a fire to escape from a field or [from] where permission for burning has been obtained
4. Cultivation of fields in existing rainforest is forbidden
5. Felling or damage to trees on hilltops is forbidden
6. Every householder must plant and care for ten trees per year for the next 6 years when supplied with seedlings
7. Felling of trees anywhere without permission is forbidden
8. Ridges cultivated on sloping ground must follow the contour
9. Where ridges are not planned, they as well as storm drains must be provided
10. Destruction of vegetation within 20 yards of a stream is forbidden

Today, regulations 1, 2, and 3 are broken with impunity since enforcement and proof of guilt is nearly impossible to obtain. Regulation 7 is completely ignored as *nkomanjila* is made wherever possible. Ridges are more common today, but more the result of the general process of agricultural intensification than in deference to the law. In essence, sets of regulations such as those listed sought to stabilize the traditional systems and decrease their assumed destructiveness. It has become obvious in recent years that these systems must evolve in the face of population growth and cash cropping, both of which lower the effective food-crop land available per capita. Hence, in terms of environmental protection, forest reserves have taken the place of ubiquitously promulgated but unenforceable agricultural regulations. Agricultural extension activities have attempted to support the ongoing process of agricultural intensification.

Forest Reserves

The need for conservation of forests on inselbergs in Africa was stressed by Clement Gillman in his presidential address to the South African Geographical Society (Gillman 1938). The jointed, bush-covered slopes of these erosion remnants conserve moisture and provide springs. The Mbozi planters also realized this fact and were instrumental in the setting aside of forest preserves in which human intrusion was illegal. As a result of the Mbozi Development Plan, the first forest reserves were demarcated followed by several others (see front end papers). All but the Ivuna Forest Reserve are protective and are areas of soil erosion risk as well as water catchment. The two Ivuna reserves are for productive purposes. Limited cutting of valuable *Pterocarpus angolensis* (Kiswahili, *muninga*), *Afzelia* sp. (Kiswahili, *mkora*), and *Stercurlias* sp. (Kiswahili, *mpalamusi*) as well as *Brachystegia* (Kiswahili, *miombo*) is permitted by license. The Ivuna, Isalalo, and Chumwa reserves are directly controlled by the Forest Division of the Ministry of Agriculture, Forests, and Wildlife. The remaining reserves are Local Authority Reserves, operated by the District Council. Problems of illegal encroachment on the forest reserves is greatest in the latter type. The Mbeya District Forester, Mr. F. Metze (1967), stressed the importance of demarcating additional reserves in Mbozi before the population becomes so great that removal of families from proposed reserves becomes costly if not impossible. The national goal is to have 8% of the land in forests. Mbozi now has less than 160 out of 3700 square miles set aside, some 4%. Forest reserves are analogous to cash cropping in their effective raising of population density on food-crop land resources with resulting agricultural intensification.

Agricultural Extensions[6]

The British encouraged African commercial production in the period before World War II, but most expenditures were made for administration rather than development. Much of the agriculture ministry's early efforts compelled rural change with high priority on erosion control measures and cultivation of such crops as cassava for famine relief. Routine, simple innovations instituted by agricultural officers were beginning, however, to have an effect (Ruthenberg 1964). Greater efforts toward development after the war brought the Groundnut Scheme, a massive failure (Wood 1959). The decade of the 1950s saw a large increase in production from both African and European holdings. Rising prices in the early 1950s combined with larger fund allocations to agriculture accelerated the development impetus.

[6]For a discussion of official agricultural policy in Tanzania see Ruthenberg (1964).

It was only with the opening of new and cheaper transportation links to the major coastal markets that food-crop production on a cash basis would be profitable. In 1945 as the Lupa goldfield market was diminishing, it was obvious that local foodstuff surpluses could not stand the transportation cost to the coast or the central railroad line. In 1948, Southern Highlands maize cost more per bag in Dar es Salaam than Kenya maize. Hence, in the postwar period the government sought to introduce or encourage cash crops able to stand high shipping costs (Tanzania Archives File 11039; Fuggles-Couchman 1964).

By the middle 1950s the British had evolved an agricultural development policy for improvement of existing peasant farms through extension, institution of voluntary schemes, and organization of sales through cooperatives and marketing boards. The major objective of extension was to persuade farmers to grow more cash crops. A number of innovations were introduced, including new crops, crop varieties, and insecticides. These were spread by extension officers using demonstration plots and persuasion (Makosya 1965). Extension officers were to work cooperatively with community development workers in increasing the people's willingness to change. Efforts were often concentrated where most needed or where the results might be most significant. In Mbozi, such a concentration in the Focal Point Approach (1956) was attempted in which extension was concentrated in the Vwawa, Igamba, and Ruanda areas. The decade of the 1950s was also noted for the rapid growth of cooperatives at government encouragement. The preindependence development plan (1961–1964) followed recommendations of the International Bank (1961) in placing agrarian efforts ahead of industrialization.

Shortly after independence the regional, area, and village development committees were instituted (Chapter 2), with agricultural extension officers to serve as advisors (Ruthenberg 1964:114–116). In Mbozi *bwana shambas* have continued to act as purposeful agents of change. However, as in other areas of Tanzania, self-help schemes and cooperatives have become increasingly important.

In 1967 the agricultural extension group in Mbozi Area consisted of 24 field staff who, in general, lived in rural areas and operated small demonstration plots for their own family consumption. Extension methods used included day-to-day advisory work on continuous tours through assigned areas; talks at village meetings; demonstrations on coffee farms; and displays at local festivals such as independence day and commemoration of the founding of TANU.

Much of the Mbozi extension work is with improved cash-crop production, although attention is devoted to problems in the traditional sphere as well. In 1967 efforts toward encouraging soil conservation practices in hilly areas were relatively futile as was the introduction of DDT to control the maize stem borer. Perhaps the latter was fortunate in view of the widespread disillusionment with DDT in many areas of the world. Efforts toward increasing cassava production as a famine measure in Kamsamba were encouraging. Farmers who had been hiring tractors every year were encouraged to buy plows so that they would be independent. Demonstrations in the use and upkeep of plows were held (Mbozi Agricultural Office 1968).

In 1967, the Mbozi agricultural extension staff seemed to be a vital group. They had much to offer regarding new cash crops and techniques for cash-crop production, but some of their efforts toward improving food-crop production met with little success. Their advice regarding early planting was ill-advised, and for other improvements they lack ideas and techniques to "extend." Each staff member has 1100 families in his area, an unfavorable ratio where other means of communication of agricultural information are not available. Hopefully, future experimentation at Mbimba in new crop varieties and husbandry methods will provide viable improvements in the traditional food-crop sphere.

Up to the present time, agricultural extension has not been as important in the spread of cash-crop production as the estates. However, as the estates decline in importance, Nhiha will rely increasingly on *bwana shamba* for guidance. Hence, in the future even the agricultural extension efforts will, through emphasis on cash crops and new techniques, continue the process of change and intensification extant in traditional Nyiha agriculture and accelerated by the missionary, planter, and trader.

Models of Change

chapter 7

"Ingombe inzoni zitakuwa niziva."–
A bellowing cow has no milk.
NYIHA PROVERB (Busse 1960:131)

We have now identified a series of elementary components within the complex reality of Mbozi. These elements are the people; their models of the environment; traditional and evolved agricultural systems operating within a man-modified landscape; population and economic growth; and sources of change. First, we must order these isolated features of modern Mbozi. For this ordering, four basic perspectives allow us to place Mbozi, the particular, within more generally applicable approaches of our science. A cultural perspective views the evolution of man–environment systems through time as a development of thought, skills, and technology resulting from local initiative in response to population pressure on resources. An ecological perspective focuses on crucial issues of environmental productivity and the capacity of environment to sustain human population through specific exploitation systems. An economic perspective

197

draws our attention to the cumulative processes of socioeconomic growth and increasing envelopment of local production systems within regional, national, and global economic processes. Finally, the evolution of spatial organization, encapsulated in monitoring the geographical pattern of change through time, provides a crucial perspective based on a fundamental geographical principle—that interaction in space increases with propinquity. Exploration of the application of these four perspectives in Mbozi is our first task.

Our second concern will be the reassembly of the isolated elements and processes characterizing Mbozi. We will again turn to a general approach within which Mbozi can be placed, the geographical pattern of modernization. The modernization model describes spatially the interaction of elements tied together by the development process in order to discern patterns of development at a given point in time. Groups of covarying elements indicate major dimensions of modernization; when we turn these element groups into scales for a specific time and measure locales along them, we portray the spatial status of the modernization process at that time. The particular modernization process in Mbozi will be illustrated as a dynamic model of change, assembled from the four basic perspectives of analysis.

The Cultural Model

Possible correlations between population densities and resource characteristics have long interested geographers, anthropologists, and others. There exists no universal model linking population and resource potentials. This is due at minimum to questions of time scales and cultural evolution; to multiplicity of examples of dense populations in seemingly low potential environments, sparse populations in seemingly high potential environments, and former dense populations no longer existent; and to the difficulty of controlling other such critical variables as historical and military developments, invention and diffusion of production systems, and constancy of resources. Thus, approaches to population/resource questions may come from either direction–population or resources. From the population viewpoint, recent work by Boserup (1965) has captured much current thought.

Boserup's basic premise is that population growth is an independent variable determining agricultural development. Successive stages in the evolution of agricultural systems represent increasing frequency of cultivating land necessitated by increasing population density. These steps have correlates in increasingly complex tools; an initial increase in labor requirements for a given yield; integration of livestock with arable farming; increasingly dense settlement and transportation patterns; increasingly complex social infrastructure, sociopolitical organization, and division of labor; increasing dependence on a money economy; and increasing tenacity of land tenure (Boserup 1965). The model is applicable primarily to premodern societies in which economic incentives have not short-circuited the population-intensification process.

Underlying this perspective is a focus upon a people receiving or devising more intensive agricultural systems to cope with exigencies of population growth. Either local innovation or diffusion of production systems from elsewhere provide potential solutions to scarcity of land with respect to population. My interpretation is that within the society these alternative skills and technologies are evaluated by cognitive processes; conditioned by social and political systems; and eventually become incorporated as part of the ongoing cultural package of cognition and behavior with respect to the environment. Among the important considerations in the transition to more intensive agricultural systems is a cognizance of greater productivity from land evolving from, and perhaps eventually replacing, concern with returns to labor. Boserup proposes five stages of development reflecting increasing intensity of land use.

Forest fallow or shifting cultivation is characteristic of societies with low population densities and a low level of technological development. As population grows, only secondary bush can develop during the shorter *bush fallow,* and often land is kept in cultivation for longer periods. Improvement of tools and agricultural techniques makes *short* or *grass fallow* cultivation possible, with only a year or two of fallow following an equal period of cultivation. The animal-drawn plow becomes important for cultivating the grass fallow. Greatly improved fertility-maintaining techniques enable *annual cropping* to support high population densities. Finally, as population continues to grow, *multi-cropping* takes place. The latter two stages can develop along the capital-intensive Western model with migration away from the agricultural sector, or follow the Oriental model of agricultural *involution* in which the increasing population is absorbed in the rural economy. [1]

The Unyiha landscape reflects beautifully an ongoing process of agricultural intensification and change. Cash cropping and delimitation of forest reserves have been analogous to population growth in initiating the use of more intensive, grass fallow agricultural techniques. In this way, Unyiha parallels the model proposed by Boserup. At low population densities, the *nkomanjila* woodland system of cultivation is prevalent. When repeated clearing, shorter fallows, and longer crop sequences were developed in response to limitation of land resources, the fallow landscape became a cultivation steppe. From a matrix of slash-and-burn fields and regrowth bush and woodland, the landscape has become a mosaic of fields and fallow grassland punctuated with fruit-bearing trees. Grassland systems of cultivation became predominant in upland, agriculturally derived grassland. The introduction of cattle and plough was contemporaneous with this intensification process, completing the transition to the short fallow stage proposed by Boserup. The necessity for more food and compensation for more labor was at least partially met by increasing emphasis on cassava production.

The same degree of agricultural intensification is yet to be realized among the Namwanga and Wanda, although techniques more intensive than bush fallowing do occur near villages and on the margin of natural grassland areas. General population densities in these areas are lower than Unyiha, and cash cropping is less prevalent.

While the Boserup thesis does not fully apply in Mbozi because of vast economic changes accompanying population growth, it is clear that increasing population densities here, as elsewhere, may not solely result from a larger food supply, but food supply mustered to support population increases related to medical,

---

[1]Involution refers to an agricultural system capable of absorbing additional labor by producing at least as much output as needed to support the added labor input. Rice cultivation systems of Asia fit this description (Geertz 1963).

biological, economic, and political factors. Hence, the historical perspective of Boserup provides a more satisfactory explanation of increasing population densities than would potentiality of land resources.

Models developed for Mbozi from the Boserup perspective can be expressed graphically and cartographically as well as verbally. A directed graph (Figure 7.1) indicates the role of population growth, cash cropping, allocation of land to estates, delimitation of forest reserves, and environmental modification (particularly, deforestation) in altering the available land for food-crop production per person. A number of alternative reactions to this pressure on resources occurred in Unyiha—intensification of production in the form of crop rotations, root crop production, and plowing; internal migration; and an increasing tenacity of land tenure, as illustrated by the Silwimba farm discussed earlier. In addition, an increased population led to environmental alteration and the substitution of grassland agricultural techniques for those used to manage a wooded fallow. Finally, intensification (the ubiquitous response in Mbozi) or controlled population growth (only incipient, with changing attitudes toward large families and polygyny) create a balance between population and local resources. Do note that the graph (Figure 7.1) is not Boserup's model, but appears to be consonant with her argument.

Cartographically, the idealized spatial development of agricultural intensification begins with a small region of forest fallow *(nkomanjila)* agriculture supporting a traditional core area (Figure 7.2, part I). As population grows through time, outward expansion supports larger numbers with the same system (Figure 7.2, part II), an expansion that eventually reaches the borders of the region (Figure 7.2, parts III and IV). In the meanwhile, shorter lengths of fallow in the core area result in bush fallow, then short or grassland fallow *(nkule, itindiga)* agriculture becoming predominant, with each increase in agricultural intensity radiating outward from the core as population growth and internal migration maintain the population/resource balance. The last map of the sequence reflects in very abstract form the situation in Mbozi in the 1960s, and indirectly the visible landscape—forest

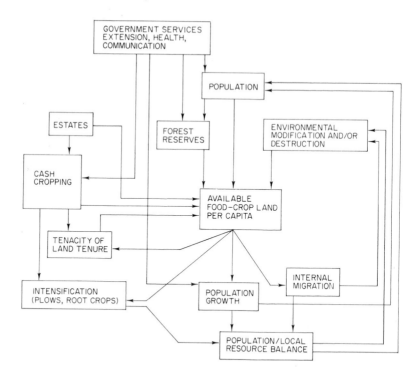

Figure 7.1. A cultural model of agricultural change.

fallow/cropland mosaic in the periphery and highly intensive, grassland fallow agriculture with emergent annual cropping on the man-created cultivation steppe in the core area.[2]

The Resource Model

The other side of the population–resource argument is, of course, the resources needed to support a given population. Although we can generally dismiss much of the recent and

[2]Note the resemblance of this cartographic model to the now familiar von Thünen model of annular agricultural zonation based upon economic rent accruing to land at successive distances and therefore transportation costs from a central market town (Chisholm 1967). Economic rent is not really operating here except in the limited sense of accessibility to the social and political core of the region. Rather, we would expect the concentric formulation to disappear as internal migration equilibrates population densities. However, when we add economic development to our argument, von Thünen rings will appear, reestablishing the concentric pattern.

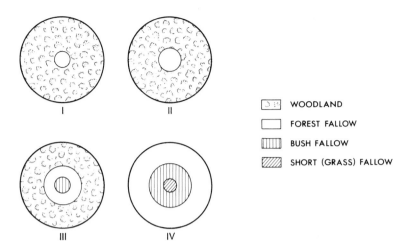

Figure 7.2. A cartographic model of agricultural intensification.

rational argument questioning the potential productivity of tropical areas as only partially applicable to highland, rather temperate Mbozi,[3] one necessary perspective on change in Mbozi is
provided from the environmental resource perspective—the
combined arguments of William Allan and Afolabi Ojo.

Allan's argument begins with a classification scheme in which
the population-carrying capacity of an agricultural system is
determined from resource characteristics. To estimate this figure,
it is necessary to have a soil- or land-type map; an estimate
of the proportion of the land cultivable in each soil- or land-type
category; an estimate of the duration of cultivation and fallow
for each category; and an estimate of the land area currently
in cultivation for each person (Allan 1965:8–9). Given these data,
Allan derives a land-use factor, the total number of fields that
must be in various stages of production or fallow for each field
presently in cultivation. For example, if a field is cropped for
2 years and rested for 10 years, the land-use factor is 6. Based
on the land-use factor, Allan classified land types as follows:

1. Permanent cultivation land: land use factor less than 2
2. Semipermanent land: 2–3

[3]Lee (1957), Hodder (1968), Chang (1968a, b), and Grigg (1970) have discussed
environmental limitations on tropical agricultural development. Highland Mbozi
has a more favorable potential photosynthetic status than lowland tropics, however. In combination with the better Mbozi soils, this suggests fewer environmental limitations here than elsewhere in the tropics.

3. Recurrent cultivation land: 4–10
4. Shifting cultivation land: 10 or more

The cultivation factor is the amount of land presently in production per person. From the land-use and cultivation factors, one can calculate the Critical Density of Population (CDP) of an agricultural system, "the maximum population density the system is capable of supporting permanently in that environment *without damage to the land* [Allan 1965:89; his emphasis]."

Allan's system accepts current agricultural techniques as a premise from which the CDP of particular land resources can be calculated. Once the agricultural system is given, the limitations are then imposed by the environment. In analyzing areas of high population density in East Africa, Allan suggests that these areas, on the whole, have soils of relatively high fertility. However, he does grant that such factors as population growth and cash cropping can bring about an alteration of traditional land-use systems. Allan's general contention is that population growth is critical in Africa. His view of the future of agriculture and the resources of Africa is less than optimistic: "Without . . . leadership, reinforced by technical assistance and financial aid on a larger scale than has hitherto been available, Africa may remain, as it has been described in the past, 'A riddle without an answer' [Allan 1965:474]."[4]

Allan's discussion of African agricultural systems and the resources supporting them has been amplified by Ojo (1968), who has suggested factors that alter the Critical Density of Population. Those that lower the CDP include:

1. Private land tenure that makes land less fluid in the society
2. Relatively permanent destruction of soil resources due to past agricultural practices

[4]A typical view of shifting cultivation that parallels that of Allan is Yudelman's (1964:13):

> . . . under the system of shifting cultivation, a relatively sparse population was in ecological balance with its environment. The problem today is that the environment has changed. Population has increased, and, in some instances, European occupation has limited the land supply available for shifting cultivation by African producers. Nevertheless, many of the traditional methods of production have persisted, taking a very heavy toll of the soil.

3. Increased acreages needed per person under modern mechanized farming methods
4. Introduction of cash crops that take land from the traditional farming system
5. Higher health standards which, because of greater labor effort available, have increased the amount of land cultivated per person
6. Improving skills such that many educated people who return to rural areas want larger holdings

Factors that raise the CDP, according to Ojo (1968) are:[5]

7. Scientific farming practices
8. Indirect population pressure of economic migrations of people to other sectors of the economy
9. Direct reduction of population pressure by migration to town
10. Freeing of land formerly off-limits for agriculture

We may sum the Allan–Ojo model in this way: Given a static agricultural system, there exists a critical population density which, if exceeded, will initiate a spiral of increasing infertility of the natural resources. However, there are a number of factors that will alter the critical density. These factors and their relative importance will vary with particular circumstances. Hence, when the only information available is that population is growing, the long-term result is equivocal.

The central tenets of the Allan–Ojo argument can also be illustrated graphically (Figure 7.3). Population growth and the removal of land from potential use (for example, delimitation of forest reserves) both result in less available food-crop land per capita. Unless population growth is checked or internal migration takes place, under a given agricultural system the length of fallow must decrease; regeneration of fertility during fallow will lessen; more land will be required to feed each person, fur-

[5]In a wider regional context, arguments 8 and 9 are spurious. Unless migrants purchase food from abroad, the overall population that must be supported remains the same. Only if migration serves to optimize local population pressure and/or productivity systems can it be said to raise the CDP. Frequently, urban migration is accompanied by a demand for higher quality foods, less ecologically efficient to produce. As requirements for animal protein increase, the CDP must decrease, reflecting the loss of plant material as it is cycled through animals.

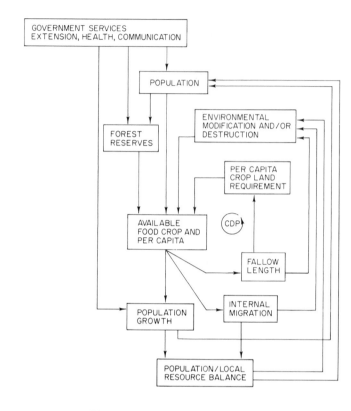

Figure 7.3. A resource model of agricultural change.

ther decreasing the fallow land per person and resulting period of fallow; this results finally in a collapsing spiral, perhaps eventual disaster. The Mbozi experience is, in the short run, that a Boserup kind of response has avoided the major detrimental consequences of this model. However, we did note reports of an increasing size of field required to feed one person signifying at least a partial working of a portion of the Allan cycle.

From a similar environmental perspective, we know that resources are not evenly distributed within Mbozi area, but that a marked gradient occurs from the hilly southeastern area with ample rainfall and rich soils to the northwest with its low and unreliable rainfall and poor soils. Realizing this environmental distinction is not the sole determinant of agricultural practice, we can cartographically suggest the way in which environmental character has skewed the simple intensification model presented earlier (Figure 7.4). Here the core area is less centrally located, but expansion of the successive rings of intensification has been limited by the hill land on the periphery and the less desirable environment in the northwest, although eventually all of the area is occupied by forest fallow agriculture (Figure 7.4, part IV) if not by more intensive systems.

The resource perspective thus indicates potential difficulties that could result should agricultural intensification not follow population growth; at minimum it requires us to skew our analysis in the direction of resource quality gradients.

The Economic Model

Clearly, the development process in Mbozi is embedded within and dependent upon a national and international world of markets and money. Money serves as the local translator between produced goods and services; it also ties the area and its products into the larger systems, providing the means by which Mbozi is able to import from national and international markets a myriad of market commodities both desired and necessary. A voluminous literature has accumulated on the role of agriculture in

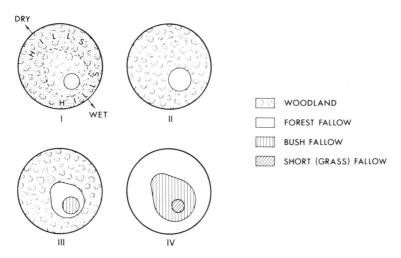

Figure 7.4. A resource-skewed model of agricultural intensification.

national economic development—in Tanzania in particular.[6] Our analysis has been at a totally different scale of resolution, however, with specific focus on household production and the local community. Thus, it seems reasonable to build a simple, locally applicable model from the Mbozi experience and earlier studies.

In his study of modernization in Kenya, Soja (1968:4) listed factors that contribute to the extent of development in an area:

1. Development of the transportation network and increased traffic on it
2. Expansion of information and communication media
3. Growing rural–urban interaction
4. Breakdown of ethnic dichotomizations
5. Growth of the money economy
6. Development of education
7. Participation in extralocal organizations and activities
8. Proximity to and interaction with modernized segments of the society
9. Degree of mobility within and beyond the local area

Many of the factors are particular manifestations of the general concept of increased accessibility. Specific features that have increased accessibility in Mbozi include the bicycle, radio, roads, and potentially the railroad.

Accessibility may simply mean going to centers of activity, to town (Figure 7.5). Most Nyiha travel to Vwawa, Itaka, or Mlowo several times per year. Only a few go to Mbeya, a 50-mile journey on the twice-daily bus. For those who do go, a trip to this town of about 12,000 people is a major event. The town, like the *duka* in the bush, is both a purveyor and creator of desires. The sale of one's produce or labor to achieve certain desires creates an increasing dependence on the market economy. At the same time, it creates further steps of achievement—clothing, bicycles, motorcycles, automobiles. Bicycles and radios are particular foci for large-scale purchases in Mbozi. The bicycle increases accessibility, resulting in continued exposure to new ideas—both for making money (new crops or techniques of pro-

[6]For example, see Johnston 1960, 1967; Eicher and Witt 1964, 1968; Baker 1965; Mellor 1967; Southworth 1967; Gaitskell 1968; McPherson 1968; Mosher 1966; Makings 1967; de Wilde 1967. Specific discussion within a Tanzanian context has been provided by Ruthenberg 1964 and Smith 1965.

Figure 7.5. The symbols of change. The gentleman was employed in the household of a European resident of Mbozi. His clothing was purchased at the Mbeya market and his bicycle at the COSATA (government trading firm) store in Mbeya.

duction) and spending it. At the same time it may make entering the labor market possible since travel time to a distant job becomes reasonable. On a larger scale, cash cropping within Mbozi may help justify improved linkages to other areas.

In addition to accessibility and concomitant contact with new objects and ideas, taxation may play an important initial role in the transition to a money economy. In his study of the Bwamba of Uganda, Winter (1955) proposed a model of agriculture change demonstrating the effect of taxes as an economic incentive and alternative development processes resulting in a local cash farming or industrial economy (Figure 7.6). The impetus to change is provided by imposition of taxation. In most cases, taxes have been a means to secure revenue for government operations. In many cases, taxation was also intended to force local people to provide labor for expatriate estates or industries. Hence, a limited number of cash crops or seasonal labor migration provides revenues sufficient to meet tax demands. A number of

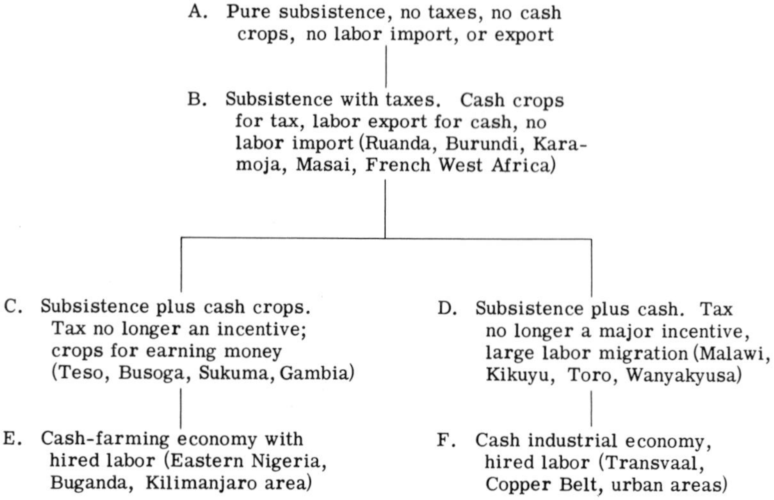

A. Pure subsistence, no taxes, no cash
crops, no labor import, or export

B. Subsistence with taxes. Cash crops
for tax, labor export for cash, no
labor import (Ruanda, Burundi, Kara-
moja, Masai, French West Africa)

C. Subsistence plus cash crops.
Tax no longer an incentive;
crops for earning money
(Teso, Busoga, Sukuma, Gambia)

D. Subsistence plus cash. Tax
no longer a major incentive,
large labor migration (Malawi,
Kikuyu, Toro, Wanyakyusa)

E. Cash-farming economy with
hired labor (Eastern Nigeria,
Buganda, Kilimanjaro area)

F. Cash industrial economy,
hired labor (Transvaal,
Copper Belt, urban areas)

Figure 7.6. Stages in the transition from simple subsistence economies to cash economies. [Adopted from Winter (1955).]

other factors including new desires, increasing dependence on purchased goods, and the possibility of rapidly accumulating wealth (wives, cattle) that previously would have taken years to accumulate became the major determinants of development. Continued development can follow two lines, toward a cash-farming economy or toward an industrial economy with migrant labor.

Although taxation was indeed an imposition forcing people into the money economy, much of the development in Unyiha goes far beyond requirements for meeting tax obligations. Virtually all of this growth has been voluntary, Nyiha responding to opportunities and becoming increasingly involved in the modernization process. This process can be graphically presented (Figure 7.7). Whether taxation, school fees, increased material desires, or increased dependence upon the market for formerly family produced commodities motivate further commitment to the modern economy, three alternative sources for income are available. These include cash cropping, craft activity, or entering the labor pool for possible employment on coffee estates—in the service of craftsmen, in commercial establishments, or through labor migration outside the region or even nation. Most of these income-producing activities provide an additional reinforcement of increasing economic desires and market dependence. A significant dimension of increased wealth may be use of food imported to the region—cattle from Msangano, fish from Lake Rukwa, flour from Dar es Salaam, canned and frozen foodstuffs from Kenya and abroad.

Increased wealth and accessibility support further investment of government services, which in turn provide increased school opportunities, agricultural extension, and growth of the transportation and communication linkages that rebound to increase accessibility even further. Thus, economic growth in Mbozi has apparently achieved a kind of "take-off" resulting from the compounding of modernizing processes, a take-off that promises continued growth should the national and international milieux remain hospitable.

The pattern of economic growth has to a very large extent followed the spatial and temporal spread of cash-crop production, especially cultivation of coffee. Focusing on the fourth of our perspectives, we view the process of modernization through space and time; then we will reassemble Mbozi into contemporary reality.

The Spatial Model

Our fourth and final perspective focuses on the process of development or modernization as it is manifested through space

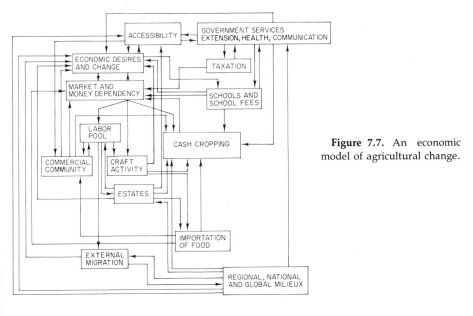

Figure 7.7. An economic model of agricultural change.

and time. During the 1967 field session, the local political hierarchy (Figure 2.11) served as a means to disseminate and collect carefully constructed, simple questionnaires in Kiswahili for every *Kumi-kumi* group of households in Mbozi Area (Figure 7.8). Because of the difficulty in completing survey research in a largely preliterate society, the basic questionnaire used (Table 7.1) had to be concise yet academically incisive. Thus in addition to obvious questions concerning population size, labor migration, and occurrence of cash cropping, bicycles, radios, and plow (all indicators of local modernization), the dates of first occurrence of each of four critical items (coffee, plow, bicycle, radio) were also queried. The latter provided a time series of data that allowed exploration of geographical patterns of change through time, and virtually into the future. The questionnaire achieved results totally unexpected—over 60% returns, with some 1300 *Kumi-kumi* groups replying (Knight 1969).[7]

Figure 7.8. The *Kumi-kumi* survey. Here, questionaires are being passed out to *Kumi-kumi* chairmen at a Village Development Committee meeting at Chiwanda in Unamwanga. Note the mango tree in the background.

[7]Unfortunately, no returns in the *Kumi-kumi* survey were received from Msia, Nsenjele, and Wasa VDCs (see Appendix 2). Field reconnaissance, agricultural surveys (Wasa VDC), and the commercial surveys completed in these areas provide a basis for estimated data shown on maps.

Table 7.1

The Kumi-kumi Questionnaire

Mbozi, 1969

Kwa Bwana Kumi- kumi:

Salaam. Tafadhali tunahitaji msaada wako katika kitabu chetu cha sehemu ya Mbozi.

Mkuu wako, mtendaji mdogo amekubali kutusaidia kwa kupata ruhusa ya kuwauliza maswali ya fuatayo; kwa msaada wa viongozi wa kumi-kumi na watu wa kumi-kumi. Tafadhali andika majibu ya maswali na ukijibu tafadhali mpatie mtendaji wako mdogo karatasi yenye majibu haraka iwezekanavyo.   Asante sana.

  1. *Jina la unapokaa: V.D.C.* _____ *Mahali halisi* _____
  2. *Kuna watu wangapi katika watu wa kumi-kumi wako?* _____
  3. *Watu wangapi wa kumi-kumi wako wana baiskeli?* _____
  4. *Mwaka gani mtu wa kwanza katika sehemu yako alinunua baiskeli yake?* _____
  5. *Watu wangapi wa kumi-kumi wako wana radio?* _____
  6. *Mwaka gani mtu wa kwanza katika sehemu yako alinunua radio yake?* _____
  7. *Watu wangapi wa kumi-kumi wako wana jembe la ng'ombe?* _____
  8. *Mwaka gani mtu wa kwanza alinunua jembe la ng'ombe lake?* _____
  9. *Watu wangapi wa kumi-kumi wako wanalima kahawa?* _____
10. *Mtu wa kwanza kulima kahawa alianza mwaka gani?* _____
11. *Watu wangapi wanalima pareto (maua)?* _____
12. *Watu wangapi wanalima ngano?* _____
13. *Watu wangapi wanalima paddy (mpunga)?* _____
14. *Watu wangapi wanalima simsim (ufuta)?* _____
15. *Watu wangapi wamekwenda kazini katika sehemu nyingine?* _____

In the discussion that follows, each *Kumi-kumi* group will be considered as a single unit. It was suspected and subsequently confirmed that at the desired level of aggregation for analysis—the Village Development Committee—a strong relationship would exist among the following types of variables: the time since initiation of a modernization variable (for example, coffee growing), the number of individual families having that variable at a given time, and the percentage of *Kumi-kumi* groups having at least one family growing coffee at the same time. As discussed later, this relationship reflects an elegant regularity in the diffusion of modernizing innovations in space and time. In addition, compelling reasons existed for minimizing the length and complexity of the form sent to every *Kumi-kumi* cell. Many cell leaders were illiterate, requiring them to seek the aid of school children or neighbors in actually filling out the document. Because the leader required consultation with members of his group, I felt that the less complex his task the more likely he was to complete it well. Related to this argument is the fact that bicycles and radios, critical indicators of development in Mbozi, are both subject to annual government taxes which are not always paid. Thus identity of individual owners and even of *Kumi-kumi* groups had to be protected. Since the simple questionnaire asked no questions that would positively identify the group and only aggregate totals of items were requested, it seemed reasonable that more reliable totals of taxable items would be forthcoming. To avoid compromising the whole community of informants, may I simply suggest that this supposition was critically correct, with the survey reporting more radios and bicycles, for example, than annually registered at Area headquarters with tax paid. Finally, it seemed reasonable to suspect that in the aggregate, regularities of modernization would appear within the *Kumi-kumi* group as well as among groups. This hunch also proved accurate, based upon data gathered from a much more lengthy questionnaire attached to every fifth of the simple questionnaires. This detailed sample requested both the simple form plus a listing of household and production variables for each family, providing a basis for both confirming the *Kumi-kumi* as a suitable minimal unit for analytic discussion and checking the general degree of accuracy with which simple forms were executed. Data gathered in this survey can be analyzed graphically and mathematically.

Taking Iyula Division (see front end papers) as an example, if we plot over time the cumulative proportion of *Kumi-kumi* groups that have at least one family cultivating coffee or owning a plow, bicycle, or radio (Figure 7.9), we can see a progression of development through time. Here, adoption of plows follows an almost identical rate of acceptance to that of coffee, with a lag of approximately 5 years. It may not be coincidence that this is about the length of time it takes newly planted coffee to achieve significant yields. Purchase of bicycles and radios follow at a considerable lag behind the initial commitment to the major means for purchasing them—cash cropping of coffee or surplus arable crops supported by use of the plow. To the actual point data generated from the survey we can statistically fit lines representing the best approximations to the data we have gathered. Note that these best-fit lines can be extended beyond the time of data gathering (1967) into the future.

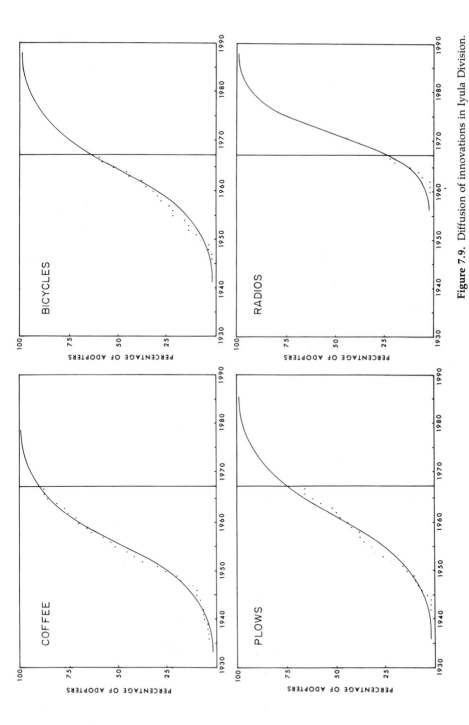

Figure 7.9. Diffusion of innovations in Iyula Division.

215

Similar graphs were created for each individual Village Development Committee and Division as well as Mbozi Area as a whole (Figure 7.10). For Mbozi, the upper limit to coffee adoption is considered to be 70% of the aggregate number of *Kumi-kumi* groups in the area, representing Unyiha, the rest of Mbozi being ecologically unsuited to coffee. For plows, bicycles, and radios the asymptote remained at 100%.

The general form of the curve of innovation acceptance is one in which the proportion of adopters is small at first, increases slowly, then rapidly rises, and finally decreases in growth rate as the total proportion of potential adopters is approached. It can be shown (Casetti 1969) that when the following basic premises are applicable, acceptance of innovations must assume a curve of this form:

1. Acceptance of the innovation results from potential adopters learning from adopters.
2. Potential adopters are resistant in varying degrees to acceptance of the innovation.
3. Within any area, there are a variety of potential adopters with different degrees of resistance.
4. Individual resistance to accepting the innovation is surmounted by repeated contact with adopters.

It should be clear from earlier discussion that the diffusion of coffee cultivation in Mbozi precisely fits these premises. Thus, we would expect the spatial and temporal pattern of coffee cultivation in Mbozi to fit nicely within more general diffusion models. It is likely that the diffusion of plows also fits the premises, while that of radios and bicycles may more logicallly be viewed as assuming a similar form based partially on the premises and partially upon the income derived from coffee or plows that facilitated their purchase. These diffusion dynamics were multistage processes. All Nyiha—adopters and nonadopters—know of coffee growing as they certainly did in the past. The spread of this knowledge of the *possibility* of coffee cultivation was virtually instantaneous compared with the slower process of becoming acquainted with cultivation techniques in sufficient detail to undertake cultivation oneself. In addition, Brock (1966 and personal communication) has suggested that the earliest cultivators were subject to both Nyiha ridicule and European opposition. The early innovators required even more than know-

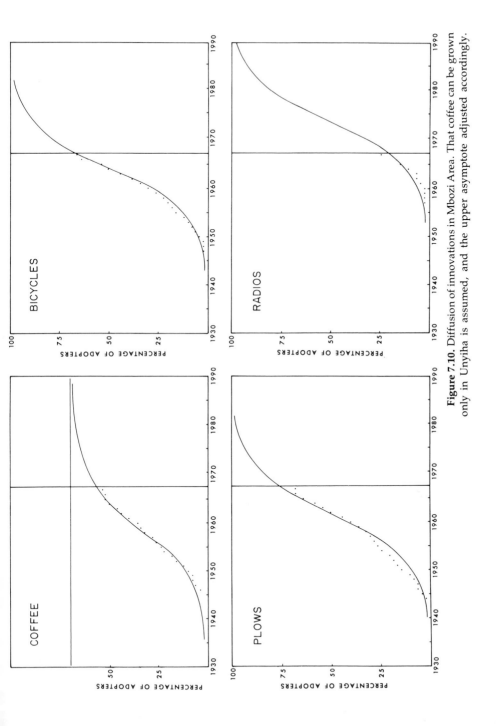

Figure 7.10. Diffusion of innovations in Mbozi Area. That coffee can be grown only in Unyiha is assumed, and the upper asymptote adjusted accordingly.

217

ledge—the willingness to tolerate unfavorable reactions and perhaps to find social identification that supported their deviation from current norms. This commitment, ability, or willingness to change lagged considerably behind simple awareness of the innovation.

It can be further shown that the logistic curve is one among an infinite number that could result from the four diffusion premises (Casetti 1969). This logistic form of the acceptance curve specifically implies a contagious type of diffusion—requiring spatial contiguity between potential adopters and those who have accepted the innovation—with an adoption rate at a point in time proportional to the current percentages of potential and actual adopters (Brown 1971:552). Beal and Bohlen (1957), Rogers (1962), and others have identified various portions of the logistic curve with adopter dispositions—innovators and early adopters represented in the early and slowly increasing portion of the curve, the majority identified during the most rapid phase of diffusion, and laggards or nonadopters characterizing the final portion of the curve in which it slowly approaches the upper asymptote.

Modeling the diffusion of innovations in this way is particularly attractive because it allows us to use statistical methods to find the particular logistic curve parameters that best fit a given time series of diffusion data. The basic form of the logistic curve is

$$P = \frac{U}{1 - e^{A-B}}$$

where P equals the proportion of adopters at time T; U represents the upper asymptote or the total potential adopters (usually but not always 1.0); A is a parameter indicating the relative position of the logistic curve in time; and B indicates the diffusion rate or the relative shape of the curve (Gould 1969). A low A value places the curve earlier in time, and a low B value indicates a curve of slower diffusion. Using this mathematical formulation, we can fit curves to the actual diffusion data gathered in Mbozi.[8]

[8]Logarithmic transformation of the data allowed use of a standard regression program to fit a line; the data and fitted line were transformed back to initial form; and the fit of the curve was improved by iteration. *Rho* values were calculated from the iterated curves.

That the mathematical model represents actual VDC, division, and areawide data well is indicated by *rho* values uniformly ranging upward from 0.70, with the majority of curves for coffee, plows, bicycles, and radios having *rho* greater than 0.90. *Rho* can be interpreted as the proportion of variation in the dependent variable (percentage of adopters) that is accounted for by the fitted curve. The resulting curves may be used in several ways.

First, the curves indicate a significant regularity of acceptance of the innovations through space and time. Thus, we are not examining a happenstance or random pattern of change. Rather, the modernization indicated by the spread of coffee and plows, bicycles and radios is an ordered process of acceptance of change and the spread of this commitment in a regular fashion across the region.

Second, the fitted curves provide a basis for examining the nature of the acceptance itself. One might expect that as coffee cultivation spread into such ecologically less suitable areas as the margins of Unyiha, the rapidity of acceptance of coffee cultivation would have waned (Figure 7.11A). The Mbozi experience is quite the opposite, however, with more marginal areas having considerably more rapid acceptance curves once coffee growing was initiated (Figure 7.11B). The basic parameters of the logistic curve provide a means for comparing the time of initiation of coffee growing with the rate of its acceptance (Figure 7.12). The significant positive relationship between these parameters in

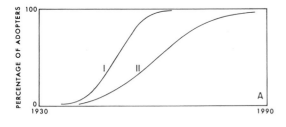

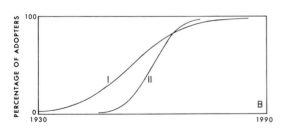

Figure 7.11. Adoption of coffee cultivation. A—the ideal case with most rapid diffusion in areas most ecologically suitable (I) and less rapid diffusion in more marginal areas (II). B—the actual case in Mbozi with later, more marginal areas (II) having a more rapid adoption pattern.

Unyiha suggests a very strong internal impetus for agricultural change, with marginally located peoples generating their own desire to move into the market economy. This is, of course, obvious from either a mechanistic or humanistic viewpoint. In the former it is clear that at a later date of initiation the number of previous adapters in adjacent areas from whom a potential adapter can learn is larger; hence, a faster diffusion process. Humanistically, people in peripheral areas will have been aware for some time that the market economy is desirable and that means to enter are available. Once these means become locally available they are accepted with little hesitation.

The market commitment in Unyiha was rapidly followed by purchase of commodities such as bicycles and radios, establishment of local services (butchers, grain mills, restaurants), and establishment of local general stores, all of which furthered rural development. The spatial distribution of all of these features, both presently and through recent decades, reflects the fundamental, underlying diffusion of market orientation in production and most specifically of coffee cultivation. The burst of rural modernization in the 1950s was paralleled by the widespread establishment of Local Authority schools during the period from 1952 to 1964. The TAPA schools (Chapter 2) in more peripheral areas seem to have resulted from further demands for education once the modernization impetus was realized there.[9] Outside Unyiha, in neighboring cultural areas of Mbozi, the diffusion of the plow serves as a similar indicator of the commitment to development, although the results of market orientation are presently less visible than in Unyiha.

The strength of the local initiative in agricultural change is demonstrated by the very rapid acceptance of innovations in production once sufficient detailed knowledge becomes available. The existence of strong local initiative should not be surprising, given the equally strong ability of local peoples to reject and resist development schemes that are economically or socially unacceptable.

Finally, the fitted logistic curves may be used as predictive

[9]The compressed period in which the Local Authority schools were founded indicates no clear diffusion pattern. Unfortunately, the Mbozi Area Education Office could not provide records of the founding dates of the TAPA schools.

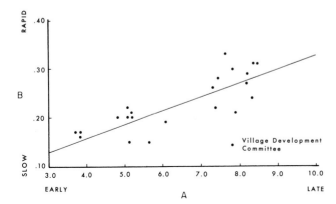

Figure 7.12 Date and rate of acceptance of coffee cultivation. The A and B values are parameters of the logistic curve. A indicates the relative date of acceptance and B the rate of diffusion. Note that villages with late initial acceptance of coffee have more rapid acceptance once the diffusion has begun. The fitted line has $R^2 = 0.69$. Each dot represents one village.

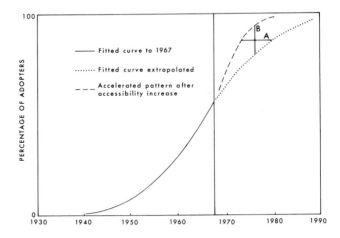

Figure 7.13. The potential impact of increased accessibility. Acceleration of the diffusion of innovation is suggested by the change in curve trajectory. The impact could be measured in either time (A) or quantity (B).

221

devices. Note that curves presented here all extend past the actual date of data gathering, 1967, with predictions as extensions of an ongoing process. Now consider the probable impact of improvements of the Great North Road and construction of the Tan–Zam railroad. We would expect these transportation developments to rapidly increase the accessibility of Mbozi within the national and international context, thus accelerating rural development. Measuring this impact may be more vexing. It would certainly be erroneous to attribute all change since 1967 to transportation improvements. Thus, using the curves as predictors of patterns likely to occur in the future with a continuation of present growth processes, deviations related to transportation changes can be measured, properly attributing to transportation only that acceleration beyond what would have occurred (Figure 7.13).

The logistic curves indicate through time a process moving across space as well. We have seen from earlier discussion and maps of coffee cultivation at three different time periods (Figure 6.9) that a marked regularity of adoption occurs over the years, and moreover that coffee cultivation has radiated outward in space from a core area of European coffee estates. This radiation is also implied in discussion of the regular way in which areas late to initiate coffee cultivation also experience a more rapid pattern of adoption (Figure 7.12). An idealized spatial spread of coffee cultivation can be cartographically presented (Figure 7.14). European coffee estates and delimitation of a local road network initiate a process by which coffee cultivation spreads from the estate to African small holding, with the spread pulled slightly along the existing roads because of increased interaction along them. This pattern essentially encapsulates the local modernization process in Mbozi. The spread of plows followed a very similar pattern, but moved beyond the bounds of Unyiha to incorporate most of Mbozi Area as a means to enter the market economy with surplus crop production (Figure 6.11). The spread of bicycles and radios followed the general spread of the means to attain them—coffee and plow.

When we bring all of these perspectives on change—cultural, resource, economic, and spatial—up to contemporary time, we then reassemble Mbozi into reality.

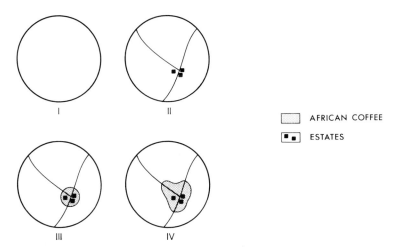

☐ AFRICAN COFFEE

▪▪ ESTATES

Figure 7.14. A cartographic model of coffee diffusion.

Models of Change

We are now able to reassemble Mbozi by merging our perspectives on change. To accomplish this, we begin by unifying the graphic models of the change process created in our discussion from the cultural, resource, and economic perspectives. Then, a cartographic model will be used to illustrate this process of change as it has been experienced in Mbozi in recent decades, drawing from the idealized maps illustrating change from the cultural, resource, and spatial perspectives. Finally, we use data collected as indicators of change in Mbozi to build a numerical and cartographic portrait of her patterns of modernization.

Merging the three separate graphic models of change (Figure 7.15), linkages tying them together focus upon government services, cash cropping, land tenure, agricultural intensification, migration, and perhaps most critically the available food-crop land per capita. Local change is, of course, embedded within government activity and the more general regional, national, and global milieux, whose influences literally and figuratively surround and enclose the process of modernization in Mbozi. Time may be the most significant unspecified parameter, especially relative timing among the various subsystems in the model. In Mbozi, economic growth has occurred concurrently with the

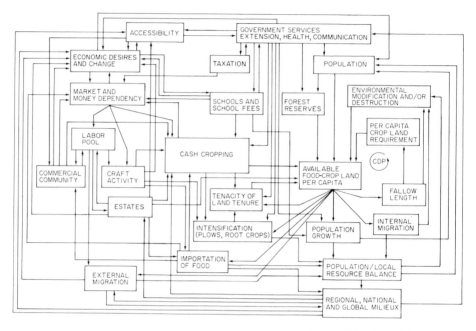

Figure 7.15. A model of change.

process of agricultural intensification, and adequate land resources have been available during both transitions to allow evolution toward economically, culturally, and ecologically viable land-use systems. However, one can easily imagine circumstances in which the motivation to involvement in the cash economy occurs late in comparison to the population growth/intensification process, severely restricting the availability of resources for cash-crop production without massive inputs of capital and knowledge from outside the local system. Similarly, economic growth could precede any land shortage, perhaps accelerating modernization as technology becomes available—the European occupation of rural North America being a prime example.

The synchronous pattern of change in Mbozi evinces several emergent characteristics. First, agricultural intensification as a result of population pressure on resources, allocation of land to cash crops, and decreased fluidity of land resources due to more tenacious land tenure is reflected in the transition to a grassland fallow agricultural system, with the increasing importance of the plow as an agricultural implement. Second, prime among alternative sources of income to meet both increasing economic needs and desires is cash cropping, reflected in patterns of land use, the evolution of land tenure, and in commodities and services that income has made available—schools, bicycles, radios, *dukas*, grain mills. Finally, this dual pattern of agricultural intensification and economic modernization has a spatial dimension, written by the broad paintbrush of resource characteristics and temporal development of change from critical foci and accelerated along routes of accessibility.

In idealized cartographic form, change in Mbozi (Figure 7.16) reflects the diffusion of coffee cultivation from the region of European estates. In addition, the idealized map reflects the acceleration of the resource-skewed, culture-derived intensification process as land is allocated to other than food crops, with a larger proportion of land in more intensive agricultural systems than might otherwise have occurred at this time (Figures 7.2 and 7.4). The concomitant spatial distribution of commercial activity (Figure 7.17) suggests the role of cash cropping coffee and increased production of food crops in supporting commercial growth and increasing dependence upon the market for numerous commodities and services. If superimposed, the two maps would reflect the idealized spatial pattern of modernization in Mbozi—a core area of coffee cultivation and intense involvement in the market economy and commerce; peripheral areas of only

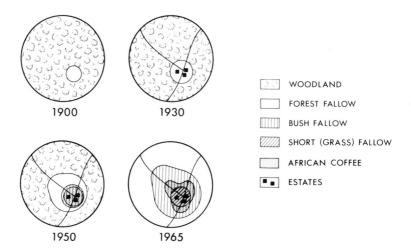

WOODLAND

FOREST FALLOW

BUSH FALLOW

SHORT (GRASS) FALLOW

AFRICAN COFFEE

ESTATES

Figure 7.16. A cartographic model of agricultural change.

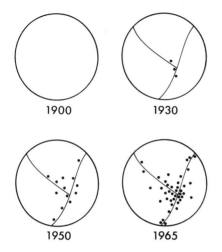

1900   1930

1950   1965

Figure 7.17. A cartographic model of business establishment distribution.

slight involvement, awaiting perhaps the impetus of newly found accessibility or viable and profitable new cash crops. But rather than superimposing these maps of the ideal, let us turn to the pattern of the real.

Rural Modernization

The two surveys mentioned earlier, one directed to individual *Kumi-kumi* groups through the Village Development Committee, the other surveying the commercial activity of Mbozi Area (Figure 6.12), provide a body of selected data that indicates the basic spatial pattern of rural modernization in Mbozi. Variables used for developing the modernization pattern include bicycles; radios; plows; coffee, pyrethrum, wheat, and sesame cultivation; labor migrants; stores or *dukas*; grain mills; bars and restaurants; butcheries; African-owned motor vehicles; and European estates, included in their role as a major source of employment (Table 7.2). It is readily apparent from the divisional data that Unyiha, with its coffee, cultivation, bicycles, radios, and commercial activity is clearly the most modernized portion of Mbozi Area. In general, the absence of cash crops means a greater rate of labor migration, highest among the Namwanga.

In order to "see" this data as a whole, imagine the full data set of 18 variables creating a "space" of 18 dimensions in which each Village Development Committee is located as a point. The

Table 7.2

Mbozi Modernization Data Summary[a]

| Variables | Divisional Summary | | | | | | | Variable loading on first principal component |
| | Unyiha | | Vwawa | Unamwanga | | Uwanda Kamsamba | Mbozi Area | |
	Iyula	Igamba		Msangano	Ndalambo			
1. Bicycles per *Kumi-kumi*	1.11	1.81	1.56	0.95	1.33	1.52	1.46	+ .58
2. Percentage of *Kumi-kumi* with bicycles	0.61	0.77	0.66	0.51	0.54	0.60	0.66	+ .65
3. Radios per *Kumi-kumi*	0.37	0.57	0.21	0.29	0.30	0.30	0.37	+ .54
4. Percentage of *Kumi-kumi* with radios	0.22	0.37	0.15	0.23	0.15	0.24	0.24	+ .56
5. Plows per *Kumi-kumi*	1.26	2.68	0.49	5.72	1.05	4.93	2.44	– .40
6. Percentage of *Kumi-kumi* with plows	0.66	0.89	0.29	0.96	0.47	0.96	0.69	– .14
7. Coffee growers per *Kumi-kumi*	6.56	6.89	3.76	—	0.10	—	3.74	+ .80
8. Percentage of *Kumi-kumi* with coffee cultivation	0.88	0.91	0.66	—	0.06	—	0.54	+ .79
9. Pyrethrum growers per *Kumi-kumi*	1.22	—	0.97	—	—	—	0.42	+ .02
10. Wheat growers per *Kumi-kumi*	1.53	0.99	0.32	—	—	—	0.57	+ .47
11. Sesame growers per *Kumi-kumi*	—	0.01	0.01	2.87	—	3.88	0.89	– .29
12. Labor migrants per *Kumi-kumi*	0.88	0.79	0.83	2.04	1.56	0.83	1.01	– .38
13. Stores (*dukas*)	42	84	116	11	19	13	285	+ .78
14. Grain mills	5	12	13	1	—	2	33	+ .73
15. Bars and restaurants	3	5	18	—	—	8	34	+ .42
16. Butchers	5	17	15	—	—	3	40	+ .61
17. African-owned motor vehicles	5	4	20	—	1	3	33	+ .41
18. European estates	1	10	29	—	—	—	40	+ .30

[a]Data are from the Mbozi Economic Survey (Appendix 2) and principal component analysis.

exact location of each of the 41 village points for which a full data set exists is determined by measuring along each variable axis the distance of that village datum on the variable; so measuring for all 18 variables locates the village in 18-space. Think of all 41 village points as forming a mass in 18-space. In some directions it will be stretched longitudinally; in others it may be more spherical. If we could actually see the mass, we would find that the points were not scattered randomly, but that some strong relationships between the variables restricted it in certain directions. It would be helpful indeed to find that line or group of lines in 18-space which best characterizes our data set. We could then collapse the data set on that line or lines and reduce the dimensions of our data space to a manageable number.

Principal components analysis allows us to discern a new set of combined variables that incorporates much of the information contained in the former singular variables.[10] For example, in Unyiha we have already mentioned the strong relationship between coffee cultivation and other indices of modernization—bicycles, radios, commercial activity. A computer program was used to discern the principal components in this data set, and it derived a new set of variables, the first four accounting, respectively, for 29, 18, 13, and 6% of the variance in the original data set. Because the fourth new variable accounted for only a shade more than any single one of the original variables, only the first three components were used in subsequent analysis, and they account together for three-fifths (60%) of the total variance in our data set. Although we have now lost two-fifths of the details in our data, we have cut the dimensions of our data space from 18 to 3, a reasonable sacrifice in detail for manageability.

Our analysis can move in two directions. We may use the technique of factor analysis to help us clarify the variables we have derived, and will do so subsequently, or we may directly interpret the principal components analysis. Taking the first component, we find that the positive end of that new dimension of measurement indicates many bicycles, radios, coffee cultivators, stores, grain mills, and butcheries, that congerie of mod-

---

[10]For discussion of principal components and factor analysis, see Harman 1960 and King 1969. Computer programs used were part of the Biomedical series (Dixon 1970), BMD01M and BMD03M.

Table 7.3

Modernization Scores[a]

Village Development Committee	Score on first principal component	Village Development Committee	Score on first principal component
26 Chiwezi	-3.20	3 Mahenje	0.02
29 Chilulumo	-3.19	23 Isandula	0.08
30 Nsanzya	-3.18	41 Kamsamba	0.45
28 Chitete	-3.04	20 Ihanda	1.23
35 Namchinka	-3.00	13 Nambinzo	1.60
39 Itamba	-2.58	1 Ruanda	1.61
34 Chipanda	-2.38	22 Hanseketwa	1.67
33 Luashyo	-2.26	12 Itaka	1.72
40 Mkulwe	-2.02	11 Itandula	1.76
37 Ilengo	-1.99	7 Isansa	1.90
24 Ipapa	-1.98	25 Tunduma	2.09
38 Masanyita	-1.83	2 Iyula	2.20
27 Msangano	-1.76	4 Mlangali	2.33
19 Ihowanje	-1.67	6 Wanishe	2.65
36 Ivuna	-1.61	16 Kigoma	3.17
10 Magamba	-1.39	9 Halungu	3.18
32 Ndalambo	-1.13	15 Nsala	3.42
31 Chiwanda	-1.03	5 Igamba	4.11
18 Hezya	-0.63	14 Mbimba	5.57
17 Nyimbili	-0.57	— Nsenjele	No data
21 Ukwile	-0.33	— Msia	No data
8 Shikwale	0.02	— Wasa	No data

[a]Village Development Committee Scores on the first principal component of the modernization data set derived from Biomedical Program BMD01M (Dixon 1970).

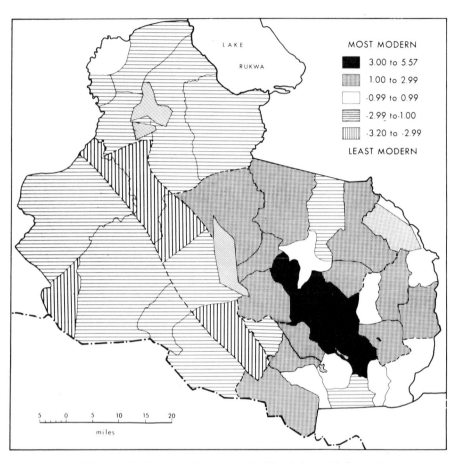

Figure 7.18. Rural modernization in Mbozi Area. Modernization scores on the first principal component are mapped.

ern activity emphasized so frequently in our discussion (Table 7.2). We can measure or project the location of each Village Development Committee along this primary component (Table 7.3), ranging from Chiwezi in the Msangano Division of Unamwaga as least modern to Mbimba in Unyiha as the most modern. Mapping these scores on our first component (Figure 7.18) provides a cartographic image of the major dimension of modernization in Mbozi. The pattern clearly follows the idealized modernization surface we proposed earlier. Although the second and third components could be handled in the same way, factor analysis provides a clearer picture of additional dimensions in our modernization pattern.

Using a factor analytic approach allows us to find for our data set a specified number of different components, termed "factors," which clarify our interpretation. With this approach a set of three factors, for example, can be derived which simplify the way in which variables are combined in forming them. Using this technique we discern three factors—the coffee–rural modernization complex; alternative income sources; and commercialization (Table 7.4). The loadings of each of the original 18 variables on each factor is used in suggesting arbitrary labels for the factors (Table 7.5). The general coffee–rural modernization factor (Figure 7.19) indicates those areas of Unyiha in which coffee, bicycles, plows, and radios form a well-knit complex. The alternative income factor (Figure 7.20) separates remaining areas not strongly involved in the first factor according to income, ranging from areas where income is derived from coffee, wheat, pyrethrum, and European estates to areas at the other end of the scale with plows, sesame, and labor migration. The third factor, commercialization, distinguishes the urban commercial core of Unyiha around Vwawa, the border crossing town at Tunduma, and the minor trading center at Kamsamba (Figure 7.21). These three individual factors are useful in inter-

Table 7.4

Modernization Factors[a]

Factor I			Factor II
Coffee–Rural Modernization Complex			*Alternative Income Sources*
Coffee	Plows		Coffee
Plows	Sesame	versus	Pyrethrum
Radios	Labor migration		Wheat
Bicycles			European estates

Factor III

Commercialization

Stores
Grain mills
Bars and restaurants
Butchers
African-owned motor vehicles

[a]Factors extracted from the modernization data set using Biomedical Program BMD03M (Dixon 1970).

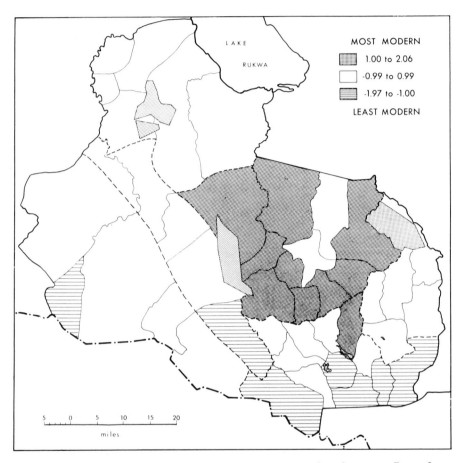

Figure 7.19. Coffee–rural modernization complex. Scores on Factor I are mapped.

preting the basic modernization pattern provided earlier in our principal component solution.

The basic modernization pattern in Mbozi (Figure 7.18) then, is related to the development following coffee cultivation in Unyiha, with Tunduma and Kamsamba villages standing out as more modern in comparison with their surrounding areas. Intermediate are villages peripheral in Unyiha, Unamwanga, and Uwanda. Least modern are Msangano Trough villages located away from Msangano itself and Namchinka village in the Ndalambo area, all in Unamwanga. The role of accessibility

Table 7.5

Modernization Factor Loadings[a]

	Factor loadings		
Variables	I	II	III
1. Bicycles per *Kumi-kumi*	**.63**	.15	.20
2. Percentage of *Kumi-kumi* with bicycles	**.76**	.19	.15
3. Radios per *Kumi-kumi*	**.87**	-.02	.06
4. Percentage of *Kumi-kumi* with radios	**.91**	-.00	.02
5. Plows per *Kumi-kumi*	.35	**-.73**	-.32
6. Percentage of *Kumi-kumi* with plows	**.56**	**-.49**	-.34
7. Coffee growers per *Kumi-kumi*	**.59**	**-.72**	.02
8. Percentage of *Kumi-kumi* with coffee cultivation	**.46**	**-.81**	.03
9. Pyrethrum growers per *Kumi-kumi*	-.28	**.43**	-.13
10. Wheat growers per *Kumi-kumi*	.17	**.62**	-.00
11. Sesame growers per *Kumi-kumi*	.11	**-.49**	-.12
12. Labor migrants per *Kumi-kumi*	-.18	**-.45**	.00
13. Stores (*dukas*)	.22	**.43**	**.75**
14. Grain mills	.39	**.40**	**.48**
15. Bars and restaurants	.06	-.13	**.90**
16. Butchers	.20	.04	**.89**
17. African-owned motor vehicles	-.08	.08	**.79**
18. European estates	.02	**.40**	.09

[a]Loadings of each of the modernization variables on three rotated factors derived from the modernization data set are indicated using Biomedical Program BMD03M with Varimax rotation and squared multiple correlation coefficients for communality estimation (Dixon 1970).

in determining this pattern is clear. All of these "least modern" villages are totally isolated. However, construction of local feeder roads in these areas could significantly increase opportunities for the Namwanga.

Mbozi illustrates an ongoing process of rural transformation resulting from population growth, allocation of land to permanent cash crops, and rural modernization. Understanding of this process may provide insight for potential instituted agricultural change or development planning elsewhere. It is also possible to indicate some interrelationships between processes and consequences of this transformation that could significantly alter the prospects of change in Mbozi.

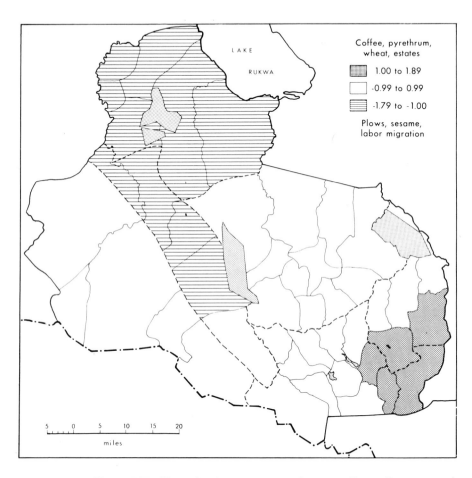

Figure 7.20. Alternative income sources. Scores on Factor II are mapped.

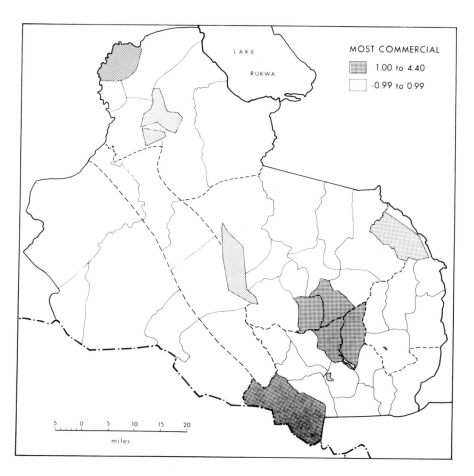

Figure 7.21. Commercialization. Scores on Factor III are mapped.

Implications of Change

chapter 8

"Timbwi tizula."–
A termite hole cannot be filled with water.
NYIHA PROVERB[1] (Busse 1960:131)

Although we cannot know the future, we can provide some glimpses beyond the present to the Nyiha and leaders of the sociopolitical system in which they live. Within a generation the Nyiha and their neighbors have made commitments toward change of a magnitude accomplished over centuries rather than decades in the Western world. Although major continuities amid change are manifest—traditional food-crop agriculture based upon a persistent ethnogeography perhaps the most striking among them—without doubt the Nyiha cannot turn back. Traditional skills that provided hoe blades, clothing, and numerous other commodities now supplied by the market economy will be lost. Communication with and knowledge of the larger world will increase; as a result, Nyiha culture and activity will adopt

[1]Nyiha proverb said when one is given a task he cannot possibly accomplish.

new ideas and objects, and will adapt to a national and global milieu. Political organization and education increasingly accelerate change which has come to pervade the innermost mind and body in the guise of world religious tradition and modern medicine. The Nyiha future will be written by forces both within their control and beyond their grasp. Awareness of some salient implications of the process of change experienced up to the late 1960s may facilitate this transition and perhaps increase the Nyiha's own role in directing their emerging destiny.

For this excursion into the land of speculation we can draw on our familiar perspectives—spatial organization, economy, ecology, and culture. Ordering of these perspectives is purposely reversed from our earlier discussion. Changes in spatial organization, economic opportunity, and human modification of the environment have all contributed to the evolution of Unyiha. However, the traditional food-crop production system still occupies the largest proportion of Nyiha time and land. Although markedly altered in emphasis among systems through the course of time, these means of basic food production—built upon traditional assessments of land resources, understandings of ecological processes, and rationale for activity—remain the dominant functional relationship between man and land among the Nyiha. Change has yet to totally transform Nyiha culture and cognition. It seems reasonable to argue that continued change will build upon traditional knowledge and communication between it and modern science. The alternative is total substitution of our system of knowledge for theirs. I strongly discourage such a substitution because I doubt both the efficacy of the educational process and the validity of our scientific system in unfamiliar environments.

Change and Spatial Organization

We have seen that the spatial arrangement of Mbozi has contributed markedly to the process of change. As we look toward the future, four geographical aspects of change warrant commentary. First, the accessibility of Mbozi within a national and international context will be altered. Then at a larger, more local scale, the way in which agricultural extension officers and other development facilities are articulated in space may have a critical impact on the pace of modernization. Land tenure constitutes a third dimension of spatial organization, one that will prove crucial as modernization progresses in Mbozi. Finally, we will attempt to assess the potential impact of Tanzanian *Ujamaa* concepts on Unyiha in view of contemporary land tenure evolution.

Accessibility

Mbozi's accessibility within local, regional, and international frameworks will be greatly extended. The Great North Road was paved in 1968–1969 in support of the Zambian oil lift. After years of preliminary surveys to develop a rail line to the southwest of Tanzania dating as far back as the German occupation, the railway has finally been constructed (Gillman 1929; Gibb 1952; Leverett 1957; O'Connor 1965; Griffiths 1968). Both transportation improvements were motivated by national and international considerations. However, these linkages will accelerate the pace of change in Mbozi as well. Additionally, a significant decrease in the cost of shipment between Mbozi and sizable markets at the Copper Belt and Dar es Salaam should encourage a wider cash cropping base. Internally, local transportation improvements will help to open the Msangano and Ndalambo areas, tapping the obvious desire for growth I observed in these areas. Thus the opportunities limited by poor accessibility in the 1960s will be considerably broadened in the 1970s and beyond.

As suggested in Chapter 7, the improvements in accessibility will have a marked effect on the pace of change. We now have both the data and a method for measuring and understanding the directions of change as they evolve.

Location of Development Agents

A second implication of spatial organization may be illustrated by the way in which the efforts of agricultural extension officers are scheduled. There is little doubt that a combination of persuasion and demonstration by the *bwana shamba* will facilitate continued agricultural change. Unfortunately, the current ratio of one officer to 1100 farmers is beyond the ideal (de Wilde 1967,1:172) suggesting that the way in which these men are used may be critical for realizing their full potential. It would seem profitable to explore means by which the evolving local political structure could be used in the agricultural extension process. Specifically, if one or two selected farms from each *Kumi-kumi* group were selected as central meeting places, *bwana shambas* could be assigned to Village Development Committees in proportion to numbers of farms and conduct extension efforts at the central *Kumi-kumi* farm on a regular basis. All neighbors could gather for these meetings, raising questions about their own farms, and the extension officer could advise the central farmer on cultivation of fields designated as local demonstration plots. Thus, *bwana shamba* would be located within the existing sociopolitical structure and function in concert with it. The same kind of spatial scheduling would be critical in the less developed areas of Unamwanga and Uwanda. Here, careful planning for extension efforts is necessitated by the relatively larger distance between *Kumi-kumi* groups in less densely populated areas. As in the more developed areas, extension efforts would capitalize on the inherent propensity of the local population to learn from itself, and would result in carefully spaced foci of agricultural development which would give simultaneous support for growth across a wide area.

Similar attention must be given to the location of schools, cooperatives, and other development facilities. Careful articulation of these centers within the spatial and social structure of Mbozi will facilitate the progress of modernization. Creation of the new district-level planning office in the period since 1967 should aid in the development of carefully coordinated planning.

Land Tenure

Access to land for cultivation constitutes the third and perhaps most critical spatial implication. The way in which a society allocates land among production units is one of the most significant systems of organization it imposes on the landscape. There is little doubt that land tenure will emerge as a paramount issue in Unyiha. The process of enclosure and *de facto* freehold of land that has evolved in recent decades requires careful examination within the content of Tanzanian national land policy. Since independence, Tanzania has rejected individual land ownership as one part of its broader development policy. Yet this is precisely the direction toward which Nyiha tenure has moved.

A combination of population growth; allocation of land to cash crops, especially those of a permanent nature such as coffee; and increasing domination of cropping techniques using only a short fallow period, if any, has been eroding the traditional system of access to land. By 1967, important changes in land tenure were evident. First, a virtual enclosure of the land had been completed in all but the most sparsely populated areas of Unyiha. This enclosure simply means that every plot of land in a neighborhood can be identified as belonging to one person or another, with no land other than the *mbuga* (seasonally flooded grassland) and rocky inselbergs unaccounted for. For most Nyiha, the land immediately around the farmstead forms a contiguous farming unit, with some scattered fields in other locales. While a considerable amount of borrowing and lending of land gave the appearance of severe land fragmentation, the pattern of agreed-upon boundaries is usually considerably simpler. This enclosure of Unyiha into individual holdings was usually accomplished through a series of *ad hoc* agreements between neighbors, although I recorded several instances of neighborhood *barazas* (meetings) to decide land boundaries. There has emerged a monetary market in land improvements, with a clear tendency toward sale of land itself, as will be shown later. Finally, the traditional process for settling land tenure disputes is in danger of breaking down. The local chiefs and headmen have been replaced by the *Kumi-kumi* structure and the Village Development Committee headed by a Village Executive Officer. The official judiciary structure is separate from the administrative structure, although *Kumi-kumi* chairmen and Village Executive Officers often arbitrate minor disputes. The important point here is that there is little legal basis for recognition of either changes from the traditional land tenure practices or sale of improvements to land, as well as land itself.

The following circumstances recorded in the field in 1967 reflect land tenure changes listed previously:

> Kapembati was born in Tukuyu. A Nyakyusa, he followed his father to the Wanishe VDC in Mbozi in 1959. He was allocated land by the local headman, married locally, and has no intentions of returning to Tukuyu. The yields here, he says, are much better than there; and,

also, it is not so crowded. Like other Nyakusa immigrants, Kapembati seems welcomed and on good terms with his Nyiha neighbors.

Daudi, a Nyiha, is 36 years old, and has married two wives. He went
to school in Zambia through standard eight and worked 3 years as an
addresser for the Zambian government. Then he returned to Mbozi and
took a job as clerk of the local court at Igamba. From 80 shillings per
month he rose to a salary of 237 shillings per month. Eventually he
saved a little over 4000 shillings from his salary. His plan was to invest this in a speculative business, make a quick profit, and then invest
in an established coffee farm and open a *duka*. He went to Lake Rukwa
and bought fresh fish which he dried. He hired a truck to go to the
Copper Belt where selling his fish he could realize a profit of as much as
four or six times his investment. Unfortunately, the fish were not properly cured. In addition, the truck had no lights so it could travel only
during the day, was delayed at the border for 24 hours, and had a
breakdown en route. A 24-hour journey took over a week by which
time the improperly cured fish were rotten. Some were salable, but
by the time Daudi returned home he had only 1000 shillings in his
pocket. This he finally invested in an established coffee farm in 1965.
With the farm he also acquired the houses of the previous owner, and
by arrangement with the local *Kumi-kumi* chairman took over the former owner's land rights as well. Daudi said firmly that land was not an
exchangeable commodity. By arbitration of the chairman and four
elders, he and his neighbors agreed to property boundaries. He was
clear in emphasizing that he had not *bought* the land from the previous
owner, but only the coffee *trees* and the houses.

Kapembati now claims one-half of a large field that bounds their
adjacent homesteads. Daudi has not cultivated the field since arriving,
but both litigants agree the field as a whole belonged to Daudi's predecessor. Kapembati, however, says he needs the land and that Daudi
has no right to it—it was not part of the purchase he made from the
other man—and anyway, Daudi is just hungry for land on which
he eventually wants to put coffee—"My needs are more immediate."
Daudi argues that since some land must always be resting, he has not
yet had a chance to bring this land into cultivation. He cites the earlier
decision that gave him the land, but now is afraid because the chairman
does not "remember" this decision. Kapembati, he angrily points out, is
married to the chairman's sister. He thinks the elders will remember,
though, and will substantiate his claim. Daudi does not hide the fact
that the land is not immediately needed for subsistence. Rather, his intention is to grow wheat here for cash sale and eventually to plant coffee in the area. Although he does not recognize it, Daudi is seeing more
clearly than others what will happen to land as more goes to coffee
without changes in the subsistence agriculture.

Nyiha land tenure in 1967 was one in which traditional rights
and practices were being bent to solve problems of land allocation
and transfer that were formerly nonexistent or unimportant.

Land tenure changes among the Nyiha in response to population growth and economic development raise some problems

that are not peculiar to Unyiha. First, it is obvious that individual tenure has evolved, but without the legal infrastructure recognizing the changes that have occurred. Litigation is likely to increase, as traditional tenure principles form a basis for undermining the *de facto* changes that are part of agricultural development in Unyiha. That disputes over traditional laws can no longer be solved in the traditional manner is cause itself for alarm. Second, the development of tenure practices in Unyiha seems to be contrary to Tanzanian government policy. Does this mean that traditional tenure is to be reestablished? Will *ad hoc* decisions continue to be made, undermining the cultivator's confidence in realizing his development efforts through the lack of any definite principles upon which he may rely? Will traditional tenure principles that are no longer viable be the means by which "the more aggressive farmer is. . .held back by the dislike and jealousy of his less successful compatriots? (Marcus 1960)". Will the situation evolve as it may from present more or less contiguous holdings to severe fragmentation requiring more drastic measures later? How will newly created families gain access to land?

In a wider perspective, this process of tenure change in Unyiha is not unique. Obol-Ochola (1969), in examining land tenure in Africa, suggested that population growth and cash cropping are the major factors resulting in tenure change. So long as land continued to be available, any disputes over land could, if necessary, be settled by opening new land. With population growth, this becomes impossible. Similarly, planting of such permanent cash crops as coffee, tea, or cocoa lent an unusual endurance to land occupation, as well as creating a clear economic value for land improvements.

A general relationship between population density and land tenure had been systematized in an "evolutionary school" of land tenure law. According to Obol-Ochola (1969) this "school" holds that there was an evolution of tenure from an initial stage of tribal ownership related to defense needs through clan and family to individual tenure with increasing pacification, governmental development, and population growth. While this theory is clearly superficial, changing land tenure under population pressure and permanent cropping was perceived by colonial authorities (Swynnerton 1966). In fact, the government of Tanganyika Territory in 1938 initiated an investigation of land tenure practices (Tanzania Archives File Acc 157 L2/11). The government foresaw that usufructory rights were likely to disappear under population growth and agricultural and change, and felt it ought to guide tenure changes into safe channels. Provincial commissioners were to collect land tenure information and forward it to the Secretariat in Dar es Salaam. One result of this effort was a paper in which the Land Tenure Advisor to the government, A. A. Oldaker, presented a summary of land tenure practices in Tanganyika. Oldaker (1957) suggested the following sequence of land tenure types in response to increasing population, cash cropping, and planting of perennial tree crops:

1. Purely communal tenure, "the use of land by all members of a local group without there being any exclusive rights in the individual, or permanent houses"
2. Permanent houses and land security granted to the individual but reversionary right to the tribe or group apply
3. " . . . full security of tenure to the individual, but without negotiability, and where reversionary rights are barely used"
4. Negotiability allowed but subject to group control
5. Negotiability allowed without control except as to tribe of transferee
6. Unrestricted individual ownership or freehold.

Oldaker noted that most Tanzanian peoples were at levels 1 to 4.

Among the more significant East African groups which have undergone changes in land tenure are the Chagga of the Mt. Kilimanjaro region and the Sukuma who live in the region south of Lake Victoria. Ruthenberg (1968) has noted that among the coffee-and-banana-growing Chagga, commercialization has led to a change toward privately owned land. Today land is bought, sold, and used as security for loans. Land rights among the Sukuma became individualized as a result of permanent cropping, population pressure, and economic development (Malcolm 1953). There was some fear that individualization among the Sukuma might lead to such undesirable circumstances as the rise of land-owning class, concentration of holdings in larger units than could be effectively occupied, and agricultural indebtedness. Already, sale of houses and other improvements suggested that sale of land might soon follow, leading Malcolm to propose the institutionalization of individual land rights with the requirement that the community act as middleman in any transactions, thus preventing some of the unwanted aspects of private tenure.

Land tenure change in Unyiha has occurred within, and will continue to be affected by, national land policy in Tanzania. As a consequence of the Treaty of Versailles, Britain was not allowed to own any land in Tanganyika. Rather, the territory's land was held in trust, and the Mandate and Trusteeship Agreement for Tanganyika required British recognition of African land law and custom. In addition, Britain was to control transfer of land rights in order to protect African interests. The Land Ordinance of 1923 declared all land to be public land, although recognition was taken of the validity of earlier legally acquired leaseholds and freeholds. Land alienated thereafter was to be held under rights of occupancy (leaseholds) lasting up to 99 years. Public land was to be administered for the common benefit and use of Africans, and any other occupancy or use required permission of the governor. The 1928 amendment to the 1923 Land Ordinance provided that the African community and the particular person occupying land were to have the same legal rights as if they held a lease from the government. This amendment did not create leaseholds or freeholds, but simply set a legal basis for the defence of traditional holdings (Chidzero 1961). As discussed, the Tanganyika government began to gather land tenure information in 1938 for the purpose of guiding land tenure change.

The East African Royal Commission (1955) devoted considerable attention to the need for land tenure change in East Africa. The Commission pointed out that traditional tenure systems provided security through the availability of new land. However, as a market economy develops "land becomes valuable as a specialized factor of production" and there is a population movement from the land to new employment opportunities in towns and cities. Lack of a money market for land prevents a potential buyer from having access to land for uses consistent with price, and prevents a potential seller from disposing of land that he cannot use effectively in relation to price. Furthermore, without the ability to buy and sell land, the Commission argued that a farmer could not specialize in production nor could he be certain whether he could sell or bequeath improvements (East African Royal Commission 1955:51):

> Those who occupy land fear that they may lose it or be deprived of it without proper compensation and with no opportunity to purchase other land in its place. Those who have no land fear that they will never be able to acquire it. Thus those who occupy land cling to it irrespective of their ability to use it properly. Those who have too little land for proper economic production on modern methods cannot easily extend their holdings, while those who have too much are not permitted to dispose of it for economic purposes at its appropriate price without consulting authorities, whose decision will not be made on the basis of economic considerations but on the basis of the old conceptions of tribal, clan or family security.

While the Commission was cognizant of the dangers of the emergence of a land-owning class, they proposed a law which would have instituted individual tenure.

In 1958 a government White Paper put forth proposals to institute freehold tenure for Africans in Tanganyika (Tanganyika Legislative Council 1958). These proposals were met with suspicion by African authorities, and were never instituted as law. Britain decided that problems of land tenure would be left to the new government at independence (Ruthenberg 1964).

Virtually on the eve of independence, the report of the International Bank (1961) on economic development in Tanganyika essentially echoed the Royal Commission emphasis on the need for individualization of land holdings. This need was predicated on questionable assertions that under existing tenure arrangements (International Bank 1961:94):

> a. Farmers have not the same interest in the care and development of their land as they would have if it were their own freehold property, and if they could not count on readily obtaining user rights in further land from the tribal area.
> b. It is difficult to introduce more intensive, planned farming.
> c. Land-use planning on an area basis is made very difficult.
> d. There is no means of preventing unproductive or destructive use of land.
> e. It is impracticable to move numbers of people from an overcrowded tribal land unit to one with land to spare.
> f. Land cannot be used as security for loans.

The mission felt that concerted action was required to establish negotiable land rights as a security for borrowing and as an inducement for improvement of land holdings. It recommended the creation of several land categories, providing for individual freeholds and leaseholds (International Bank 1961:94). Land reform and the establishment of individual titles would be handled by a land board composed of appointees representing traditional authorities, progressive farmers, and technical officers. These boards would develop rules for use of communal land and water resources, effect land consolidation where needed, and establish individual titles under an "individual tenure" system. Permanent tenure would provide for inheritance and sale within legal limits, rather than strict freehold status. The system would recognize and register individual holdings. Borrowing against future crops or the land value itself would require the use of a government intermediary in case of default. The last provision would have prevented the rise of a land-owning class through mortgage default.

It should be clear that the kinds of individualized systems of land tenure proposed through the colonial decades were predicated upon questionable ethnocentric models concerning necessary relationships between developing farmers and land. Although evolution toward individualized tenure in Mbozi has accompanied modernization, this has been an effect rather than cause of change. Traditional systems there, as elsewhere, have proved remarkably adaptable as change progressed. There is little evidence that African farmers have wantonly destroyed land resources because they were not personally owned; ownership, on the other hand, is no guarantee of good husbandry. Individual security and sale of land improvements do not require freehold tenure, as the Nyiha have proven. When land is scarce, it may be poor social policy to allow the money market to determine its allocation. Even if it is desirable for progressive farmers to accumulate larger units as their neighbors are profitably employed elsewhere in the economy, freehold tenure may be a poor implementation of this policy, leading to premature expansion and envelopment of less progressive farmers as wage laborers for their more modern neighbors. Increasing litigation over land resources requires some regularization of tenure patterns in statutory form, but not necessarily as freeholds. And if freehold or leasehold land is offered as collateral for development loans, what should become of the man struck by misfortune and unable to pay? Are there not better sources of collateral—the expected harvest improvement, for example—that do not risk permanent alienation of a man from the only resources that can support him when the ill wind passes?[2]

In 1962 the independent Tanzanian government renounced freehold tenure as a foreign conception incompatible with African custom (Fuggles-Couchman 1964). The policy, according to Segal (1968):

> conformed with the TANU election manifestos and the enunciated principles of African Socialism, based on communal rather than individual land tenure and economic development, which would not result in the emergence of social classes based on ownership of land among Africans.

The Freehold Titles (Conversion) and Government Leases Act of 1963 replaced with leases the former freeholds dating from the German period (Nasser 1965). Only one Mbozi estate, located on former mission land, was freehold at that time. This act had little effect on traditional African land tenure except as an indication of the government's commitment against the freehold system. Abolition of the former German freeholds gave little rational ground for discomfort, but it certainly contributed to the uneasiness of the European estate community.

Some of the basic premises of "African Socialism" are critical to our discussion. African socialism includes:

> reaffirmation of the values of traditional society and its collective institutions, the family and the local community; emphasis on a type of personality specific to this society; proclamation of this type of society as fundamentally different from European societies by class conflicts and as the model it can and must constitute for the new African nations [Meister 1968:158–159].

African socialism embraces both political democracy and mutual social responsibility. Nyerere (1966) has stated:

> In a country like ours, development depends primarily on the efforts and hard work of our own people, and on their enthusiam and belief

---

[2]The discussion of land tenure has been abbreviated from a paper read at the Western Association of Africanists in 1970. I was pleased to discover that Brock (1969) had independently come to similar conclusions based on her own Nyiha work and additional experience in Uganda. James (1971) has reviewed in detail the legal background of land tenure policy in Tanzania.

that they and their country will benefit from whatever they do. How could anyone expect this enthusiasm and hard work to be forthcoming if the masses see that a few individuals in the society get very rich and live in great comfort, while the majority continue for ever in abject poverty? . . . Traditionally we live as families, with individuals supporting each other and helping each other on terms of equality. We recognized that each of us had a place in the community, and this place carried with it rights to whatever food and shelter was available in return for the use of whatever abilities and energies we had.

Nyerere has further stated (Nyerere 1962; quoted by Ruthenberg 1964:132):

> . . . we must reject the individual ownership of land. To us in Africa land was always recognized as belonging to the community. Each individual within our society had a right to the use of land, because otherwise he could not earn his living and one cannot have the right to live without also having the right to some means of maintaining life. But the African's right to land was simply the right to use it, he had no other right to it . . . The TANU Government must go back to the traditional African custom of land holding. That is to say a member of society will be entitled to a piece of land on condition that he uses it.

Makings (1967) and others have proposed a kind of rightholder tenure that would fulfill minimal requirements of tenure in the context of development, yet guard against some of the major problems of individual freehold tenure. In essence, this system provides for permanent, legal, titled tenure for the individual farmer providing for inheritance, sale of improvements, and rental subject to a local land authority intermediary. Such a system would provide the continuity of tenure needed to make investment viable, eliminate any ambiguity of the spatial extent or legal conditions of tenure, and yet assure administrative control over the flow of land from person to person. Such a scheme provides an obvious and viable solution to problems of land tenure change among the Nyiha. Local authorities would assure that access to land was equitably provided. Excessive land accumulation would be controlled by a local land board. On the other hand, the system and the board would not necessarily stand in the way of accretion of holdings when it was assured that the person selling existing improvements had a viable means of support elsewhere and the buyer would be able to fully utilize the land so gained. Makings suggests that a land board could act as an intermediary in such transactions, temporarily leasing

land to the buyer where it is available, giving him the first opportunity to exchange leased land for holdings adjacent to his own
available at a later date. In this way, fragmentation of farms
as they grew in size would be avoided. No monetary market
in land as such would be instituted, and nonworking land-owner
class would be forestalled. Marked disparities in wealth could
be prevented if desired by allowing farm accretion to occur evenly
over the rural community rather than concentrated in small segments of the population. Finally, the regularization of individual
tenure does not necessarily have to affect the traditional systems
of social welfare, care of the aged, sharing of food, hospitality,
and so on. These could remain intact as local people and
authorities desire.

Rightholder tenure seems to be a potential solution to problems
of land tenure change in Unyiha that is compatible with Tanzanian land policy and a number of possible directions of future
rural development. Most importantly, this solution would provide assurance to farmers of their continued right to land already
developed, as well as land to be developed as part of their farming
plan, while avoiding the problems of total freehold tenure.
Actually, the plan simply calls for a rationalization and recognition of land tenure as it stands at the present time. Such a
procedure instituted now rather than in the future might provide
a working basis for continued development in Unyiha as well
as avoid more drastic and costly land tenure revision later.

Ujamaa

Beyond the ideology that must necessarily be linked to a discussion of land tenure, Tanzania has become increasingly com-
mitted to development within the policy of *Ujamaa*, literally *the
family* or African socialism. Essential to this policy is an egalitarian
society maintaining many of the positive values of communal
African culture and seeking development for all rather than a
selected elite. The basic commitment to this policy was outlined
in a TANU pamphlet published in 1962, elaborated in government directives, and further specified in the Arusha Declaration
of February 1967. The booklet, *Socialism and Rural Development*
(September 1967) further expanded the rural design for Tanzanian society, clarifying the nature of the *Ujamaa* villages that

had been and were being created (Nyerere 1968). While I was able to share informally Nyiha reactions to the Arusha Declaration, my departure from the field in mid-1967 (the end of a year-long agricultural cycle) precludes my reflecting personal observations of Nyiha reaction to the later, formal specification of the *Ujamaa* concept in rural areas. Fortunately, political scientist Jonathan Barker visited Unyiha in 1972, and preliminary results of his investigation of *Ujamaa* there supplement my earlier observations.

The Arusha Declaration outlined a fundamental "bill of rights" for Tanzanians; specified obligations of the party and government to the people; argued against exploitation of one man by another; emphasized the need for self-reliance in development and the critical role of agriculture; set careful limitations on the potential accumulation of affluence by TANU and government leaders; and called upon the government and people to maintain the principles of African socialism (Nyerere 1968:11–37). Nyiha reaction to the statement in early 1967 was one of enthusiasm mixed with uncertainty. It was clear that they had come a long way themselves, although considerable inequity in income among the Nyiha had already begun to emerge. As outside observers we would note, of course, that initial differences in local income are an expected consequence of modernization under a diffusion process, if not a result of differentials in resource endowment. It was unclear what the declaration would mean in specific local terms, although more than a little sympathy was directed parallel to the declaration against estate owners and the merchant community. Posters subsequently appeared with an obviously European, pipe-smoking estate owner or an equally obvious money-counting Asian merchant in the foreground, in both cases with African laborers doing heavy work in the background. The message was clear even if words had not been provided as well.

Beyond the general humane, social, and economic tenets of the Arusha Declaration, *Ujamaa* as articulated in the September 1967 booklet called for creation of cooperative villages characterized by collective farming, self-determination, and self-reliance on local issues and decisions; maximization of the impact and use of modern technology and advice; and fully socialistic lifestyles (Nyerere 1968:106–144). Since 1967 I have speculated on several consequences that would have to be weighed with respect to prospective village creation or a less drastic version of implementing *Ujamaa* in Unyiha. First, there is no clear need for formation of villages for provision of social services or agricultural facilities. Population density in Unyiha is sufficiently high that improved water supplies, equipment pools, schools and other community institutions could be provided with accessibility achievable elsewhere only by assembling people into villages. There is no obvious spatial economy in the potential formation of villages. Second, the increasing investment in long-lasting buildings and other structures on Nyiha farms would mitigate against any massive movement of households. Third, any potential change from smallholder to collectivized coffee production requires careful exploration, an experimentation that might be possible if European estate destinies were carefully planned, as discussed below. Finally, I have felt that regularization of extant land tenure changes could provide for many of the social and humanitarian aims of *Ujamaa* without radical change from present patterns.

Recent political research by Barker indicates that there are two directions in which *Ujamaa* village planning has been undertaken in Mbozi. Along the route of the Zambian railroad the central government plans a series of villages developed at existing sites, and one new site in relation to rail stations. These will include Tunduma, Vwawa, Mlowo, Songwe, and a new settlement between Tunduma and Vwawa in Mbozi Area. In 1973, these centers are at a planning stage. In addition to centrally planned *Ujamaa* developments, several *Ujamaa* villages have been organized in Unyiha under local impetus. These centers are particularly significant because they reflect both the opportunities and difficulties of instituting *Ujamaa* in developing, cash-crop economies such as Unyiha.

The most outstanding of the *Ujamaa* villages is located in Isansa. Here the village grew from the earlier conversion of the private holdings belonging to a major political figure into a cooperative (*ushirika*) farm, from which it then became a full *Ujamaa* farm. This development was paralleled in other parts of Unyiha—initial impetus by a local leader, followed by local people joining in an *Ujamaa* scheme. Often the leader provided substantial coffee holdings as a focus for the village. According to preliminary reports from Barker (1973), these *Ujamaa* farming groups are only slowly forming agglomerated settlements as existing homesteads require replacement.

The Isansa *Ujamaa* has about 60 adult members who cooperatively cultivated 17 acres of coffee, 90 of maize, 135 of beans, and 5 of wheat during 1971. This village has become well known; it received a tractor as a gift from Nyerere in 1969. Work on food-crop fields is done by work parties, thus avoiding any need for complex accounting of returns to each person. Income from cash-crop production, especially coffee, does require accounting, however, and one skilled person maintains the *Ujamaa* books. Other *Ujamaa* groups in Unyiha vary considerably in their accomplishments and *esprit*. However, they all potentially share a number of advantages. Among these are close government supervision and agricultural extension, as well as opportunity for financial aid. For women *Ujamaa* offers the help of men in the food-crop production and the opportunity to share in the proceeds of cash-crop sales (Barker 1973).

The uneven development of Nyiha *Ujamaa* groups reflects a number of difficulties. Land disputes between *Ujamaa* groups and local landholders have emerged, and relationships between the groups and the wider Nyiha community have yet to be rationalized. Barker reported considerable misunderstanding of *Ujamaa* by other Nyiha—including traditional fears of witchcraft; an erroneous belief that households, wives, and children are shared; and justifiable concern over disposition of existing coffee farms and other developments should *Ujamaa* be universally instituted. Even within *Ujamaa* groups, questions of compensating members for former investments have not been solved. Equally perplexing will be problems of providing government aid and training leaders to cope with increasing numbers of *Ujamaa* and economic diversification within them. Countering the pressures against *Ujamaa*, however, were gentle reminders from the past—elderly people reminding the youth of the communal farming heritage focused on the chiefs, and of the former village settlement pattern. Perhaps *Ujamaa* was neither new nor strange.

There is little doubt that *Ujamaa* will be strongly pressed on the Nyiha by TANU and government authorities. There is considerable opportunity for research to suggest development of

Ujamaa in directions consonant with continued productivity and with evolution of Nyiha attitudes. The strong individualism of Nyiha coffee cultivators emerged against considerable social pressure. The same desire for modernization could be effectively channeled toward *Ujamaa* with careful, equitable planning; education; and interaction among Ujamaa villages and between these pioneers and the larger community.

One salient characteristic of emerging Nyiha *Ujamaa* groups is their location in geographically peripheral areas, Isansa being an excellent example. We saw earlier how marginal areas have been highly receptive to change; indeed, this impetus having only slowly been fulfilled in smallholder agriculture may have found an alternative outlet in *Ujamaa*. It seems reasonable to hypothesize that in the more peripheral, sparsely settled areas of Mbozi — Unamwanga and Uwanda — *Ujamaa* based on full village development will be well received. People here already live in agglomerated settlements, and their clear desire to become incorporated in the modernization process has been cited. My knowledge of these peoples is too incomplete to speculate further.

Change and Economy

Among our perspectives, future economic implications of change in Mbozi are least clear. The role of the coffee estate is likely to diminish; the potential growth of the Tanzanian coffee industry is severely circumscribed by the world coffee market, thus limiting continued coffee expansion in Mbozi; the magnitude of opportunities for expansion into other cash crops is yet to be ascertained; and potential capital sources for further development are uncertain. Most critical for the Nyiha is the very fact that modernization has brought concomitant vulnerability to economic processes originating far beyond Mbozi.

Mbozi coffee estates face a future of great uncertainty, due in large part to the real or imagined threat of eventual government intervention. The fear that restrictions on taking money out of Tanzania would be tightened has been realized since 1967 and was undoubtedly an increased impetus to an accelerated departure of Europeans from Mbozi.

In terms of the Mbozi economy, were the estates no longer to exist, an important source of extra local cash income would be lost. However, from the point of view of agricultural change, the role of the estate is now past, and its functions toward continuing innovation and dissemination must be served by the Mbimba experiment station and the agricultural extension staff. Although some European estates have been purchased by Asians, it would be of great mutual benefit were the estates of those additional Europeans wishing to leave purchased by or with the aid of the government. The estates thus purchased could become cooperative farms with experimentation on a variety of management systems, and later reallocation of the farms would be possible. Past experience with abandoned estates suggests that the capital investment in coffee will be lost in the 2 or 3 years that litigation to free the leasehold might entail. Bush fires attack the coffee and pilferage threatens the buildings.

Given the opening of the paved road and especially the railroad, Mbozi's potential for supplying foodstuffs to the developing urban markets of coastal and inland Tanzania as well as the Copper Belt may be enhanced. Analysis of the role Mbozi is to play in Tanzania's contribution to the world coffee market should be made and the results translated into an effective policy of encouragement or control of coffee planting. Then, direction can be turned to other cash crops or farming systems of reasonable economic potential, and alternatives to coffee can be promulgated. For example, a thorough exploration of small-scale production of high quality beef and pork on mixed farms should be undertaken. It is critical that studies be made to suggest new direction for agricultural development in Mbozi (Figure 8.1).

One of the most acute needs for further agricultural development in Unyiha is capital (Chambers 1965). Mbozi has come amazingly far with the input of virtually no institutional credit to African smallholders. The poorest farmer is unlikely to be able to generate enough capital to accelerate his own growth process; the farmer who would most benefit from capital and serve as a source of information for his neighbors is similarly hampered in his development without it. A high priority should be put on availability of small loans against future harvests for helping subsistence farmers start toward agricultural development— for seed, fertilizers, oxen, and plows. In many cases it

Figure 8.1. The search for alternative cash crops. The late N. A. Allan, one of the original Mbozi planters, had experimented with cultivation of geraniums *(Pelargonium* spp.) whose pungent oil is used in soap and perfume manufacture. Once produced in Kenya, the distilled oil has little bulk compared to value and would offer an alternative cash crop sufficiently valuable to be shipped long distances. Unfortunately other florals have now taken the geranium's industrial role.

will require relaxation of collateral requirements or creation of new sources of funds.

Perhaps the most important economic consideration as we look toward the future is the very fact that modernization is synonymous with increasing commitment to and dependence upon an economic system far larger in scale than traditional society. This dependency, then, imposes a vulnerability upon the Nyiha to processes beyond their control. Prices and the whole realm of economic costs and returns are the primary expression of this risk. This jeopardy in the face of increasing economic individualism may be one critical factor in Nyiha receptivity to the implications of *Ujamaa* and potential growth within economic systems of greater self-determination and self-reliance.

Change and Ecology

A reasonable guideline to ecologically viable change favors those kinds of changes which simultaneously increase productivity and only minimally decrease future potential use of the environment. Thus, more intensive agricultural systems may increase productivity, but if they threaten the long-term health of the human environment through risk of accelerated erosion, for example, they could not be considered viable. The ecological perspective on future change has two dimensions, then, one of increasing productivity and the other of conserving the resource base for future use.

Increasing productivity of the landscape over a multiyear period has already been achieved by the Nyiha. Although we suggested that in a given year there is evidence that fields now must be larger to supply the food needs of a person, the greater frequency with which land is cultivated compared to the former forest–fallow agricultural systems suggests that the long-run productivity of Nyiha use of the land has increased. Nevertheless, there exists considerable room for improvement within evolved, but traditional agricultural practice. Wrigley (1961:17–18) has suggested that agricultural development might proceed with the following steps:

1. Prevention of soil erosion
2. Full use of water resources with emphasis on development of irrigation and conservation of rainfall
3. Plant breeding
4. Increased use of inorganic fertilizers
5. Protection of crops to reduce losses in the field and in storage
6. Development of animal husbandry and its integration with crop farming
7. Increased efficiency of labor

In Unyiha considerable effort is made to conserve soil with the *mandi* (ridging) system and with careful incorporation of green manures in the soil. Resources for irrigation are severely limited, suggesting more efficient use of available soil water during the rainy season. Application of experimental techniques in developing new crop varieties, both internationally at such institutions as the International Institute for Tropical Agriculture (Ibadan, Nigeria) and locally at the Mbimba Agricultural Experiment Station, should help increase use of soil water and nutrients in the form of increased crop yields. Important in the development of new, hybrid varieties will be experimentation on their utility within traditional land management systems; examination of new risks they impose with their demands on timing, fertility, moisture resources, and labor; and exploration of their acceptability as foods.

Although the potential gains to Nyiha agriculture of crop protection are not clear, it is certain that an integration of livestock with arable farming does have considerable potential here. Not only will increased use of the plow permit greater efficiency of labor application, but hopefully, rational, small-holder farming systems can be jointly evolved by the Nyiha and agricultural extension and experimentation officials. Present Nyiha landholding and land management patterns already suggest a rudimentary development in this direction, an impetus which should be capitalized upon to carry their progress forward. Although the Mbozi District Council presently owns and rents the services of a conventional four-wheel tractor, given the size of Mbozi land holdings, it is unlikely that full-scale tractors will be feasible for individual farmers in the future. Rather, an emphasis on integration of crop and livestock production should be sought. A single-axle cultivator could take the place of oxen and plow, but the required land to pay for and "feed" such a machine with no return in the form of manure suggests that it would be economically if not ecologically less viable than cattle.

Directly related to the question of agricultural productivity and continued economic modernization is the question of population with respect to resources. Although the gross population density of Unyiha in the late 1960s indicates little imminent danger of population pressure on resources, we should note that the general density figures ranging from about 8 to over 42 acres per person among Mbozi divisions (Table 2.2) hide at minimum a proportion of land unsuitable for cultivation as well as land that has been allocated to forest reserves or is at least temporarily incorporated within coffee estates. It will be crucial for a careful examination of the population/resource question to be undertaken to suggest the amount of time yet available before continuing population growth could result in a crisis situation. A general indication of the necessity for such an examination is suggested by the observation that given a minimal food-crop land (cultivated and fallow) requirement of approximately 2 acres per year per person and assuming that approximately one-fourth of the land in Vwawa Division of Unyiha is not available for cultivation, an estimated maximum population that could be accommodated without a total alteration in farming systems would be about 120,000 or slightly less than four times the 1967 population. At an annual growth rate of 3%, the population of Vwawa Division could reach that number in less than 47 years, or within the lifetime of this generation of Nyiha children! Although other areas of Mbozi have lower initial population densities, lesser resource endowment probably means that critical population densities there will be substantially less than in Vwawa Division.

Change and Culture[3]

Underlying the wide range of receptivity or resistance to change in human societies, including the Nyiha, are culture and thought. In Africa, we are caught in a duality—a history of successful agricultural development primarily resulting from individual farmers working within their slowly evolving traditional environment and responding selectively to opportunities versus the apparent intransigence of the smallholder to instituted change (de Wilde 1967, 2:221). Nyiha certainly demonstrate the degree to which traditional societies will respond to opportunities for modernization. Careful analysis of resistance to change has shown that much of it is rational, resulting from three general factors: (1) poor communication between the agent of change and the recipient peasant; (2) the agent's misunderstanding of local circumstances; or (3) the peasant's decision that a proposed change is simply not ecologically, economically, or socially viable. Realizing that we often fail

[3]The following paragraphs are adapted from Knight 1971a.

to understand the peasant's motivation, the call is made for translation of proposed changes into local terms. This seems to me to be the crux of the cultural perspective on change, the necessity for appreciating the thought of a society as a coherent, rational body of knowledge, evolved and tested through time, bequeathed as culture upon successive generations.

I have chosen here to refer to this body of knowledge as a society's ethnogeography when we specifically discuss relationships between man and his environment. Whether we use this term, folk geography, folk science, ethnoscience, ethnoecology, or others, we connote understanding of the crucial role that this culture-specific, locally applicable thought has for cultural dynamics, contact, and change. Three simple "R"'s encapsulate critical corollaries of this understanding: resources, risk, and rationality.

We often think of resources in terms of our own scientific system. We fail to recognize that peoples occupying environments, particularly environments unfamiliar to us, also have valid knowledge about nature. That knowledge upon which explorer's lives depended seems to have been abandoned by scientific explorers except in very circumscribed fields, folk medicines being the most commonplace. If we reflect that many "modern" medicines continue to be discovered from traditional sciences, can we not assume we might learn similarly about environments from folk sciences? It might appear inefficient to suggest that the Western scientist or technician initially document and understand a non-Western system of knowledge *before* instituting his own system. However, it would seem that we have accumulated enough experience to suggest that further application of hindsight to discern reasons for change having been rejected might be equally inefficient. Might we not apply the same concern with which traditional jurisprudence and land tenure have been documented to questions on the man–environment interface? It just may be that the Nyiha or others have something to teach *us* about nature.

Much of a traditional farming system seeks to minimize risk or uncertainty, the variability of factors influencing the outcome of agricultural activity which are beyond the farmer's control. In a very suggestive paper, Wharton (1968) has outlined some of the early work on rural acceptance of innovation, more importantly arguing that farmer uncertainty about the variability of innovations may explain his resistance to change. If a proposed change makes the farmer more vulnerable to risk—in reality or in his perception—he is likely to reject it. His perceptual probability structure is part of his system of knowledge. By study of the kinds of risk he faces and his perception of them, we will be better able to understand his ability to tolerate uncertainty. We may also discover means by which he is most likely to be convinced that an innovation is indeed viable.

Finally, there is ample evidence to convince us that the traditional African farmer is rational within his own cultural milieu. Kamarck (1965) has summarized some of the arguments about the inherent rationality of peasants, noting that "some agricultural schemes. . .have failed because the Africans affected worked out what would pay them best more accurately than the people who set up the scheme." In order to understand the rationality of the farmer, we must understand his agricultural activities from *his* perspective. We begin accomplishing this by looking at his ethnogeography.

We can only place our knowledge within the farmer's conception of resources, risks, and rationality by understanding his environment as he sees it. A Nyiha cannot be forced to view the world as we do. We must recognize him as an individual belonging to a society whose culture includes a viable system of knowledge. His ethnogeography is the critical linkage between himself, his behavior, and the environment from which he receives sustenance. By providing the opportunity for him to understand innovation within an accustomed framework, we recognize the most important structure to which he ties his activities: his ability to cope with the world.

The Vegetation of Unyiha

appendix 1

Key to Nyiha Uses
(L = leaf; R = roots)

10 BUILDING
    11 thatch

20 CRAFTS
    21 baskets
    22 decorative body ornaments
    23 fish floats
    24 handles for hoes, axes
    25 pottery making
    26 sap to catch birds
    27 utensils

30 FOOD
    31 fruit
    32 seasoning
    33 vegetable

263

40 MEDICINE
   41   animal ailments: k chickens
   42   external uses

421 cuts and sores	424 liniment
422 eyes	425 small-pox preven-
423 ears	tion

   43   internal uses

431 arthritis	434 fever
432 colds and respiratory	435 mental ailments,
ailments	headaches
433 dysentery	436 stomach ailments

50 POISON
   51   fish poison

60 REPELLENTS
   61   insect
   62   snake

The following list includes the scientific names of plants, the Nyiha name and a key to use of the plants. The sample numbers refer to samples taken in the field and identified by the East African Herbarium in Nairobi, Kenya. Nearly all trees are used by the Nyiha for building purposes and most grasses for hatch. Hence, in the list only plants most desirable for these two purposes are indicated as such.

Sample number	Scientific name	Shinyiha name	Nyiha uses
Trees and Bushes			
144	*Acacia brevispica*	*Uwombwe*	24
35, 133	*A. macrothyrsa*	*Inchala* (male)	424, 62
130	*A. polycantha*		
	ssp. *campylacantha*	*Uwombwe*	
157	*A. pseudofistula*	*Impuwe*	
71, 72	*A. sieberiana*		
	var. *vermoesenii*	*Ichese*	436R
42	*Afzelia quanzensis*	*Ipapa*	
57	*Albizia antunesiana*	*Iwuwa*	
16	*A. schimperiana*	*Intanga*	
2	*Allophylus griseotomentosus*	*Isitula*	
156	*Apodytes dimidiata*		
	var. *acutifolia*	*Kulavasiku*	10

Sample number	Scientific name	Shinyiha name	Nyiha uses
193	*Borreria subvulgata*	*Undinalikunyina*	421L
9	*Brachystegia sp.*	*Ilenje*	
47	*B. boehmii*	*Ing'anzo*	
56	*B. bussei*	*Insani*	
25	*B. longifolia*	*Ing'anzo*	
46, 52	*B. spiciformis*	*Ilaji*	
37	*B. utilis*	*Ibalala*	
31	*Bridelia micrantha*	*Isongamino*	10
77	*Burkea africana*	*Unakapanga*	24, 435
74, 138	*Brysocarpus orientalis*	*Inchala* (female)	
		Unahabalala	435
159	*Caessalpinia decapetala*	*Ulwanga*	
48	*Canthium lactescens*	*Igalilonji*	31
222	*Cassia floribunda*	*Unampanga*	
51	*C. singueana*	*Ipumbambuzi*	
78	*Combretum molle*	*Isumbi* (female)	
50	*Cordia africana*	*Izingati*	432
160	*Croton macrostachyus*	*Iwulugu*	
62	*Cussonia arborea*	*Ipembevunu, Inyola*	
22	*Dalbergia nitidula*	*Uluwewa*	50
66	*Dichrostachys cinera ssp. nyassana*	*Impangala*	436L
38	*Diplorhynchus condylocarpon*	*Itelembe*	433
45	*Diospyros kirkii*	*Inyumbulu*	31
182	*Dodonaea viscosa*	*Inkoli*	
3	*Ehretia sp.*	*Impumbuzya, Insugwa*	436 (31?)
15	*Ekebergia* sp.	*Iwozwo* or *Iwozyo*	424
6	*Entada abyssinica*	*Ifulo*	
67	*Erythrina abyssinica*	*Isewe*	23
36	*Faurea speciosa var. lanuginosa*	*Ivundavunda*	
23	*Ficus sp.*	*Itunduma*	26
30	*F. capensis*	*Ituwu*	31
10	*F. thonningii*	*Ivumu*	26
34, 218	*Garcinia huillensis*	*Isongwa*	31
40	*Gardenia goetzei*	*Ushihololo*	
73	*Heeria* sp. = Greenway 10273	*Ibanku*	
18	*H. reticulata*	*Ibanku*	436
76	*Hexalobus monopetalus*	*Ushizovu*	
26	*Isoberlinia angolensis var. niembaensis*	*Intonto*	
1	*Jacaranda mimosifolia*	(*Jacaranda*: introduced)	
33	*Kotschya africana*	*Intenga*	

Sample number	Scientific name	Shinyiha name	Nyiha uses
69	*Lannea schimperi*	*Unahawumba*	432R
112	*Lantana trifolia*	*Ulusongole*	
134	*Magnistipula bangweolensis*	*Itipwa*	31, 431
60	*Markhamia obtusifolia*	*Ulamba*	
178	*Maytenus senegalensis*		
7	*Memecylon mavovirens*	*Ufita*	10
27	*Monotes africanus*	*Ihahatu*	27
119	*Mucuna stans*	*Usesengwi*	
28	*Ochna schweinfurthiana*	*Isahara*	
21	*Oncoba spinosa*	*Iwambulang'oma*	
54, 55	*Parinari curatellifolia*	*Iwula*	31
43	*Pavetta crassipes*	*Unantandala*	
61	*Pericopsis angolensis*	*Iwanga*	10
68	*Philiostigma thonningii*	*Itukutu*	24
5	*Protea* sp.	*Insega*	436R, L
20	*Pseudolachnostylis maprouneifolia*	*Iputilampamba*	436R
29	*Psorospermum febrifugum* var. *febrifugum*	*Ilosyo*	
41	*Pterocarpus* sp.	*Ikula*	
4	*P. angolensis*	*Ininga*	
227	*Rauvolfia caffra*	*Isyunguti*	
180, 215	*Sapium elipticum*	*Igulukwa*	
79	*Securidaca longepedunculata*	*Umkalya, Unamutali*	436
13	*Shrebera alata*	*Imbaza*	24, 27
230	*Steganotaenia araliacea*	*Inyongapembe*	
32	*Strychnos cocculoides*	*Ishimilulwa*	31
12	*S. innocua* var. *pubescens*	*Inkominkomi*	31
63	*S. spinosa* ssp. *lokua*	*Inzolombo*	31
8	*Swartzia madagascariensis*	*Ushilonde*	24
17, 224	*Syzygium cordatum*	*Infwoomi*	10, 31
24, 64	*S. guineense*	*Insugwa, Ikwa*	31
75	*Tecomaris nyassae*	*Itwati*	
70	*Terminalia* sp. near *brachystemma*	*Ichisya* or *Isumbi* (male)	24, 436R
181	*Triumfetta annua*	*Izumba*	
53	*Uapaca kirkiana*	*Ikusu*	31
65	*U. pilosa* (first sample recorded in E. Africa)	*Impangwe*	31
49	*Vernonia* sp.	*Isowoyo*	
58	*Vitex doniana*	*Impunungu* (female)	
59	*V. mombassae*	*Impunungu* (male)	31
19	*Voacanga obtusa*	*Itungulufwanga*	
39	*Xeromphis taylori*	*Ushipolowe*	
44	*Ximenia caffra*	*Umleve*	31

Sample number	Scientific name	Shinyiha name	Nyiha uses
Grasses			
274	*Arundinella nepalensis*	*Infwenfwe*	
236	*Beckeropsis uniseta*	*Infwenfwe*	
94	*Cymbopogon* spp.	*Igoganyenzi*	61
164	*Cymbopogon densiflorus*	*Ushinyenze*	
202	*Digitaria* sp.	*Usankwe*	
114	*Eleusine africana*	*Insinjili*	21
242	*Eragrostis castellaneana*		
248	*Eulalia polyneura*		
268	*Exotheca abyssinica*	*Impilu*	
232	*Hyparrhenia* sp.	*Impilu*	
237	*Hyparrhenia* sp.	*Imwelelaminzi*	
252	*H. altissima*	*Impoli*	
264	*H. collina*	*Invunga*	
244	*H. cymbaria*		
245	*H. diplandra*	*Uhananga*	
92	*H. filipendula*	*Intyetye*	11
87	*H.* sp. near *diplandra*	*Impola*	
246	*H. lecomtei*	*Impiluimpiti*	
269	*H. lecomtei = H. newtonii*	*Impoli*	
97	*H. schimperi*	*Igonambila*	
238	*H. variabilis*	*Impilula*	
262	*Loudetia simplex*	*Unambuga*	
278	*Melinis longicauda*	*Unankohola*	
256	*M. tenuissima*		
117	*Microchloa kunthii*	*Uhalulu*	
122	*Panicum maximum*	*Upwota*	
243	*Pennisetum validum*	*Unambuga*	
90	*Phragmites* sp.?	*Itete*	10
91	*Rhynchelytrum repens*	*Unankohola*	
235	*R. subglabrum*		
121	*Setaria* sp.	*Infwenfwe*	
98, 234	*S. pallidifusca*	*Imwelaminzi*	
249	*Sorghastrum rigidifolium*		
100	*Sporobolus pyramidalis*	*Ishisinde*	22
241	*Themada triandra* var. *hispida*		
273	*Trachypogon spicatus*	*Impilula*	
253	*Urelytrum digitatum*		
Herbs			
116	*Acalypha* sp.	*Itozu*	33L
191	*Acanthaceae*		
203	*Ageratum conyzoides*		
220	*Aloe* sp.		
270	*Ascolepsis anthemiflora*		
167	*Asparagus racemosus*	*Igondolo*	

Sample number	Scientific name	Shinyiha name	Nyiha uses
104, 149	*Aspilia mossambicensis*	*Ihasyansya*	
		Ihahatu	436
85, 212	*A. pluriseta* ssp. *pluriseta*	*Ihasyansya*	
137	*Becium* sp.	*Unamawe*	434
153, 173	*Bidens pilosa*	*Imbune*	
136	*Borassus* sp.	*Ihamba*	22
86	*Borreria dibrachiata*	*Ikuwe*	
174	*Borreria dibrachiata*	*Ilungalunga*	
251	*Buchnera capitata*	*Ufita*	33L
260	*Cassia kirkii*		
	var. *guineensis*		
142	*Cephalaria pungens*	*Inguguna*	
216	*Ceropegia* sp. (may be new)		
126	*Cissampelos mucronata*	*Indulwe*	436
198, 217	*Clerodendrum discolor*	*Ivungatundu*	
81	*Commelina (benghalensis?)*	*Ihokwa*	
250	*C. neurophylla*	*Ihokwa*	
187	*Crassocephalum rubens*		
143	*Crotalaria* sp.	*Izumbamugunda*	
135	*C. lachnocarpoides*	*Ibalala*	33L
171	*Cyathula* sp.	*Ivangang'onzi*	421
93	*Cynoglossum lanceolatum*	*Ilemeng'onzi*	
83	*Cyphostemma junceum*	*Ishidugu*	32R
80	*C. kerkvoordei*	*Umchitijembe*	40k
204	*Dissotis irvingiana*		
	var. *alepestris*		
106	*Dolichos* sp.	*Unyongwe*	25
128	*Dolichos* sp.	*Ishisumbwe*	
263	*Dolichos* sp.	*Unsinsili*	51
113	*Droogmansia pteropus*		
	var. *platypus*	*Ilungalunga*	
105	*Echinops giganteus*	*Isonya*	40
176	*Embolia schimperi*	*Igalilonji*	424
158	*Eriosema verdickii*	*Imbebe*	33R
111	*Erythrocephalum longifolium*	*Ifunfunzelu*	
102	*Euphorbia cyperissioides*	*Uvimbamukole*	
272	*Fimbristylis dichotoma*	*Imwelelaminzi*	
132	*Gardenia subacaulis*	*Ushipolowe*	
162	*Gladiolus* sp.	*Ishivwilu*	
95	*Gladiolus* sp.	*Uwutindi*	
258	*Guizotia* sp.		
177	*Helichrysum glumaceum*	*Impelembo*	
172	*H. petersii*	*Iswunzwu*	
108	*Hibiscus* sp.	*Upombozye*	33L
228, 200	*Humularia drepanocephala*		
261	*Hypoxis* sp.?	*Ivigurudumu*	
209	*Impatiens assurgens*		
151	*Indigofera* sp.	*Vumamkola*	
165	*I. atriceps*		
	ssp. *glandulosissima*	*Ushiboku*	

Sample number	Scientific name	Shinyiha name	Nyiha uses
115	Inula glomerata	Infunfu	
169	Kotschya carsonii	Umlandala	436R
201	K. strigosa		
84	Laggera sp.	Unamasala	421
107	Lannea edulis	Ibumunsi	
170	Latana camara	Ulusongole	
124, 184	Leonotis mollissima	Infwonfwo	31
190	Leucas martinicensis		
214	Momordica foetida		
192	Mukia maderaspatana	Usote	33
208	Nicandra physaloides		
223	Ocimum suave		
150	Oldenlandia capensis var. pleiosepala	Iteka	434
109	Oreosyce africana	Isense	
257	Ornithogalum longibracte-atum	Unvwilu	
96	Pentas decora var. triangularis	Impangwe	
129	Physalis peruviana	Injama	31
141	Rhoicissus tridentata	Igombola	
131	Rhynchosia sp. ?	Ishisumbwe	
188	R. (clivorum?)		
211	Rubia cordifolia		
154	Scleria bulbifera	Iwola	
276	S. melanomphala	Iwola	
89	Sesamum sp.	Usambwe	33L, 423
11	S. angolense	Usambwe	33L, 425
226	Spilanthes sp.		
183, 196	Tagetes minuta	Iwingansalafu	
166	Thunbergia sp.	Ikululumbu	33
175, 179	T. lathyroides	Uhasyansya Imwela	
118, 146	Trichodesma physaloides	Undindindi Unahambuza	51
207	Triumfetta digitaria		
199	Vernonia superba		
206	Vernonia sp.		
189	Wahlengergia virgata		

Sedges

Sample number	Scientific name	Shinyiha name	Nyiha uses
99	Cyperus digitatus ssp. auricomus	Ihangaga	
266	C. distans	Indago	
168	Mariscus ferrugineoviridis	Ishilago (upland Idezye)	
127	M. sieberianus var. sieberianus	Indago	

Sample number	Scientific name	Shinyiha name	Nyiha uses
	Ferns		
267	*Dryopteris athamantica*	*Ulusengeselwa*	
103	*Nephrolepis undulata*	*Ulusengeselwa*	
145	*Pteridium aquilinum*	*Ulusengeselwa*	
	Bryophytes		
254	*Polytrichum* sp.	*Ityetye*	

The Mbozi
Economic Survey

appendix 2

During the period from May through September 1967 a detailed survey was made of farm economy and retail trade in Mbozi Area. The survey had two purposes. The first was to gather information on crops, economic wealth, and economic activity as an adjunct to the study of agricultural practices in Mbozi. The second purpose was to have a dateline of information against which future development could be measured. It was felt that the possibility of up-grading the Great North Road from all-weather *murram* to a macadam standard, the construction of the Tan–Zam Railroad, and the completion of the petroleum pipeline to Zambia started in 1967 would directly alter both the availability and price of goods, and indirectly affect economic activity through greatly increased accessibility to national and international markets.

I directed the Mbozi Economic Survey although much of the

actual survey work was executed by my assistant, Stephen
Kajula. The success achieved in certain aspects of the survey is
due to his efforts and the cooperation of local government officials and TANU officers.

The survey consisted of two parts. In the first part, Kajula
personally administered a questionnaire to owners and opera-
tors of four varieties of economic establishments in Mbozi. These
establishments were retail shops *(dukas)*, restaurants and bars
(hotelis and *bars)*, butcheries *(buchas)*, and power grinding mills
(posho mills). These four categories incorporate the largest proportion of all nonagricultural economic activity in Mbozi. Automobile fuel is sold as an adjunct to shopkeeping at only two
locations. There is one professional mechanic in Mbozi who
works outdoors from a one-room apartment. While Mbozi is
served by governmental and private transport, this activity is
centered at Mbeya, some 50 miles distant. Activities like car-
pentry, basket weaving, and tailoring outside of *dukas* do play
some part in the local economy. They are small in scale, and
usually a family is self-sufficient in products traditionally necessary for life. Specialization in iron smelting and forging of hoes,
knives, and the like has greatly decreased from what it previously must have been, since these products are now widely
available in *dukas* at moderate cost. It is argued therefore, that
the cash economy of Mbozi consists mainly of primary productivity (partially surveyed in the second part) with sales in local
food markets and through the cooperatives and these four major
categories of secondary and tertiary activity: *dukas, hotelis* and
bars, buchas, and *posho* mills.

Each establishment in these categories must buy a semiannual
or annual license from the Mbozi District Council. Each concern
on the District Council list was visited and interviewed using a
structured questionnaire. Many establishments had traded
hands, had gone out of business, or moved; in many cases the
owner was "away"; in other cases businesses were found that
were not on the "Books" and these were surveyed (in confidence) as well. Hence, there was moderately complete coverage
of all activity of this sort in Mbozi, and the location of each establishment was mapped as the basis for Figure 6.12.

The second part of the survey attempted to accomplish several
tasks. First, it was to provide a basis for estimating population
and hopefully later provide a comparison for the census results
taken in August 1967. We hoped to establish the extent to which
reasonable estimates of population could be made without a full

census. Second, the survey was designed to assess the distribution and historical diffusion of agricultural practices, mainly coffee cultivation and use of plows, and tokens of economic wealth, bicycles and radios. Finally, the survey attempted to gather selected data on the cash-crop aspects of the farm economy.

The administration of the second part of the survey depended on the political structure in Mbozi. In Figure 2.11 the regional headquarters for Mbozi are at Mbeya, and the chief representative of the central government in Mbozi is the Area Commissioner, a political appointee. The highest civil servant is the Area Secretary. Under them is the elected District Council headed by the Divisional Executive Officer, also a civil servant. Up to 1967 Mbozi did not have its own District Council, but its District Council was operated as a subcouncil of the Mbeya District Council. Plans were being completed in 1967 for the generation of an independent district council. Separate from the district council is the structure connecting the individual household with the Area and central government, at which level the TANU and government organization are closely tied. Each 10 or so households forms a *Kumi-kumi* group. Each group elects a representative to the Village Development Committee (VDC). In theory, from among the representatives to the VDC there is elected a Village Executive Officer (VEO). In practice during 1967 he was usually proposed by the VDC, but acted as a central administrative appointee with a small salary and a roughly civil service rank. Village Executive Officers are often transferred from VDC to VDC, so that elected representation in this aspect of the administrative hierarchy, at least, ends with the VDC. The Assistant Divisional Executive Officer, head of several VDCs, completes the hierarchy from household to District Council and Area Officers. In addition, parliamentary representatives are also elected and represent Mbozi directly at the Parliament in Dar es Salaam.

The TANU organization also reaches down to the VDC level with a branch secretary who is basically responsible for the annual selection of *Kumi-kumi* heads and organization of the *Kumi-kumi* structure. However, while the *Kumi-kumi* heads compose the VDC, the Branch Secretary answers directly to the TANU district office, whose functions are largely *influential* rather than direct.

The very well established administrative hierarchy, reaching down to (or up from, depending on the point of view) each household, provided an excellent means for administering questionnaires for the second part of the economic survey. At meetings of VEOs and in conversation with individual VEOs, forms to be given to each and every *Kumi-kumi* head were explained and instructions given for their completion. At subsequent meetings of the VDCs, Village Executive Officers passed out questionnaires and gave detailed instructions for their completion. *Kumi-kumi* heads who were not literate were instructed to seek the aid of school children, teachers, or members of the *Kumi-kumi* who could read. There seemed to be very little reticence to admit illiteracy; in Mbozi personal status is measured by more important factors. In many cases the Village Executive Officers themselves filled in forms from the oral answers of *Kumi-kumi* heads. The most common procedure was that *Kumi-kumi* heads took the forms back to their local area and filled them out in consultation with the *Kumi-kumi* members. The forms were then returned to the Village Executive Officer, who in turn returned the forms to the District Council Office or directly to us.

All of the *Kumi-kumi* heads received the same simple questionnaire of 15 questions concerning the *Kumi-kumi* group as a whole. To every fifth questionnaire was attached a much longer and much more complicated questionnaire asking questions of each individual head-of-household. These longer forms were to be distributed randomly; the only criterion for changing the person to whom the longer form was given was in the case of the particular *Kumi-kumi* head being illiterate. In this case the Village Executive Officer was instructed to give the longer form to the person immediately following and to give this *Kumi-kumi* head the shorter form only.

Results of both phases of the economic survey have been presented as tables and maps in support of the major themes being discussed. In only three VDCs were there no returns; in these cases the blame rests on the researchers, administrators who failed to secure the complete support of all grass-roots TANU branch secretaries (in addition to TANU district office approval of this project). Subdivisional response varied from 53 to 73%, and the total for the whole of Mbozi was 62%.

Key to Alienations

appendix 3

Alienations shown on the map of Mbozi (front end papers) are as follows:

Map Number	Land office Number	Map Number	Land office Number	Map Number	Land office Number
1	11393	17	11318	33	5073
2	9245	18	11455	34	10100
3	5814	19	10983	35	11229
4	5068	20	11471	36	11473
5	8169	21	5079	37	11639
6	7913	22	5080	38	11054
7	11247	23	5081	39	11472
8	10947	24	10152	40	4188
9	16545	25	5082	41	12026
10	11230	26	5085	42	5089
11	10958	27	4184	43	4189
12	11085	28	12006	44	5086
13	16546	29	4185	45	5085
14	1706	30	11231	46	5084
15	11760	31	14131	47	9847
16	10910	32	11039		

References

Albrecht, F. O.

1964 Natural changes in grass zonations in a Red Locust outbreak centre in the Rukwa Valley, Tanganyika, *South African Journal of Agricultural Science* 7:123–130.

Allan, W.

1931 Locusts in Northern Rhodesia, Northern Rhodesia Department of Agriculture *Annual Bulletin* **1**:6–11.

Allan, W.

1965 *The African husbandman.* New York: Barnes and Noble.

Anderson, G.

1963 *The soils of Tanganyika.* Dar es Salaam: Government Printer.

Artschwager, E.

1948 *Anatomy and morphology of the vegetative organs of Sorghum vulgare.* Washington, D.C.: Department of Agriculture *Technical Bulletin* 957.

Ashby, D. G. and R. K. Pfeiffer.

1956 Weeds—a limiting factor in tropical agriculture, *World Crops* **8**:227–229.

Bachmann, T.

1943 *Ich gab Manchen Anstoss.* Hamburg: Ludwig Appel Verlag.

Backlund, H.O.
   1956   Aspects and succession of some grassland vegetation in the Rukwa Valley, a permanent breeding area of the Red Locust, *Oikos* Supplement II.
Bagshawe, F. J.
   1930   Iringa Province, *Land Development Survey, Second Report*. Dar es Salaam: Government Printer.
Baker, T. J.
   1965   The role of agriculture in economic development. In *Agricultural development in Tanzania*, edited by H. E. Smith. Pp. 24–31. Dar es Salaam: University College Institute of Public Administration, Study 2.
Barker, J.
   1973   Personal communication.
Barnes, J. A.
   1954   *Politics in a changing society*. London: Oxford University Press.
Bartlett, H. H.
   1956   Fire, primitive agriculture, and grazing in the tropics. In *Man's role in changing the face of the earth*, edited by W. L. Thomas. Pp. 692–720. Chicago: University of Chicago Press.
Beal, G. M. and J. M. Bohlen.
   1957   The diffusion process, *Special Report* 18, Cooperative Extension Service, Iowa State University, Ames, Iowa.
Bennett, G.
   1965   Settlers and politics in Kenya, up to 1945, In *History of East Africa, II*, edited by V. Harlow and E. M. Chilver. Pp. 265–332. London and New York: Oxford University Press.
Blaut, J. M.
   1961   The ecology of tropical farming systems, *Revista Geografia* **28**:47–67.
   1967   Geographical models of imperialism, *Antipode: A Radical Journal of Geography* **2**:65–85.
Bogdan, A. V.
   1958   *A revised list of Kenya grasses*. Nairobi: Government Printer.
Bogdan, A. V. and D. J. Pratt.
   1961   *Common Acacias of Kenya*. Nairobi: Government Printer.
Boileau, F. F. R. and L. A. Wallace.
   1899   The Nyasa–Tanganyika Plateau, *Geographical Journal* **13**:577–622.
Bolhuis, G. G.
   1962   The protein content of tropical food crops, *World Crops* **14**:297–301.
Boserup, E.
   1965   *The conditions of agricultural growth*. Chicago: Aldine.
Brelsford, W. V.
   1956   *The tribes of Northern Rhodesia*. Lusaka: Government Printer.
Brock, B.
   1963   "A preliminary description of the Nyiha people of southwestern Tanganyika." Unpublished M.A. thesis. University of Leeds, England.
   1966   The Nyiha of Mbozi, *Tanzania Notes and Records* **65**:1–30.
   1968   The Nyiha (of Mbozi). In *Tanzania before 1900*, edited by A. Roberts. Pp. 59–81. Nairobi: East African Publishing House.
   1969   Customary land tenure, 'individualization' and agricultural development in Uganda, *East African Journal of Rural Development* **2**, 2:1–27.
   1967–1973   Personal communication.
Brock, B. and P. W. G. Brock.
   1965   Iron working amongst the Nyiha of southwestern Tanganyika, *South African Archeological Bulletin* **20**:97–100.

Brock, P. W. G.
1963 "The Mbozi syenite-gabbro complex of southwestern Tanganyika." Unpublished Ph.D. thesis. Research Institute of African Geology, University of Leeds, England.

Brown, L. and K. Cox.
1971 Empirical regularities in the diffusion of innovation, *Annals of the Association of American Geographers* **61**:551–559.

Bryan, W. A.
1959 *The Bantu languages of Africa.* London: International African Institute.

Bullen, F. T.
1966 Locusts and grasshoppers as pests of crops and pasture—a primarily economic approach, *Journal of Applied Ecology* **3**:147–168.

Burtt, B. D.
1942 Some East African vegetation communities, *Journal of Ecology* **30**: 65–146.
1953 *A field key to the savanna genera and species of trees, shrubs, and climbing plants of Tanganyika Territory. Part II. The species of the more important genera with general index.* Dar es Salaam. Government Printer.
1957 *A field key to the savanna genera and species of trees, shrubs, and climbing plants of Tanganyika Territory. Part I. Genera and some species.* Dar es Salaam: Government Printer.

Busse, J.
1960 *Die Sprache der Nyiha in Ostafrika.* Berlin: Akademie Verlag.

Carrier, L.
1923 *The beginnings of agriculture in America.* New York: McGraw-Hill.

Carter, G. F.
1968 Origin and diffusion of maize, *Geographical Review* **58**:492–494.

Casetti, E.
1969 Why do diffusion processes conform to logistic trends? *Geographical Analysis* **1**:101–105.

Chambers, D. V.
1965 Capital innovations for small farms. In *Agricultural development in Tanzania,* edited by H. E. Smith. Pp. 78–88. Dar es Salaam: University College Institute of Public Administration, Study 2.

Champion, J.
1963 *Le bananier.* Paris: G.-P. Maisonneuve and Larose.

Chang, J.
1965 On the study of evapotranspiration and water balance, *Erdkunde* **19**: 141–150.
1968a The agricultural potential of the humid tropics, *Geographical Review* **58**:333–361.
1968b *Climate and agriculture.* Chicago: Aldine.

Chidzero, B. T. G.
1961 *Tanganyika and international trusteeship.* London and New York: Oxford University Press.

Childs, A. H. B.
1961 *Cassava.* Tanganyika Ministry of Agriculture Bulletin 15, Dar es Salaam: Government Printer.

Chisholm, J. A.
1910 Notes on the manners and customs of the Winamwanga and Wiwa, *African Affairs* **9**:360–387.

Chisholm, M.
1967 *Rural settlement and land use: An essay in location.* New York: Wiley.

Clark, C. and M. Haswell.
  1966   *The economics of subsistence agriculture.* London: Macmillan.
Clarke, G.
  1947   Farm clearance and prehistoric farming, *Economic History Review* **17:** 45–51.
Clayton, E. S.
  1965   *Agrarian development in peasant economies.* Oxford: Pergamon Press.
Cobley, L. S.
  1956   *An introduction to the botany of tropical crops.* London: Longmans, Green.
Colby, B. N.
  1963   Folk science studies, *El Palacio* **70,** 4:5–14.
  1966   Ethnographic semantics, *Current Anthropology* **7:**3–32.
Conklin, H. C.
  1957   *Hanunoo agriculture in the Philippines.* Rome: Food and Agriculture Organization of the United Nations, Development Paper No.12
  1961   The study of shifting cultivation, *Current Anthropology* **2:**27–61.
  1967   Some aspects of ethnographic research in Ifugao, *Transactions, New York Academy of Sciences,* Ser. **2, 30:**99–121.
Dagg, M.
  1968   Major problems to be investigated in managing tropical soils. In *Conference on agricultural research priorities for economic development in Africa.* Pp. 1–13. Washington, D.C.: National Academy of Science.
Dale, I. R.
  1953   *A descriptive list of the introduced trees of Uganda.* Entebbe: Government Printer.
Dale, I. R. and P. J. Greenway.
  1961   *Kenya trees and shrubs.* Nairobi: Buchanan's Kenya Estates.
Davies, R. W.
  1952   Russia in the early Middle Ages, *Economic History Review,* Ser. 2, **5:** 116–127.
Dean, G. J. W.
  1967   Grasslands of the Rukwa Valley, *Journal of Applied Ecology* **4:**45–57.
Delf, G.
  1963   *Asians in East Africa.* London and New York: Oxford University Press.
de Schlippe, P.
  1956   *Shifting cultivation in Africa.* London: Routledge and Kegan Paul.
de Wilde, J.
  1967   *Experiences with agricultural development in tropical Africa.* Baltimore: Johns Hopkins Press.
Dixon, W. J.
  1970   *BMD biomedical computer programs.* Berkeley, California: University of California Press.
Doraswami, L. S.
  1942   The cultivation of small millets in Mysore State, *Mysore Agricultural Journal* **20:**5–9.
Driessen, J. C. B.
  1966   *Aromatic tobacco.* Dar es Salaam: Ministry of Agriculture, Forests and Wildlife.
East African Meteorological Department.
  1965   *Summary of rainfall in Tanganyika and Zanzibar for the Year 1964.* Nairobi: Meteorological Department.
East African Royal Commission.
  1955   *East African Royal Commission 1953–1955, Report.* London: H.M.S.O., Cmd. 9475.

East African Statistical Department.

1950   *African population of Tanganyika Territory 1948.* Nairobi: East African Statistical Department.

1958   *Tanganyika General African Census 1957, Tribal Analysis.* Nairobi: East African Statistical Department.

Eder, H. M

1965   Agricultural landforms in the New World: A preface. Paper presented at the 61st Annual Meeting of the Association of American Geographers, Columbus, Ohio.

Eggeling, W. J.

1947   *An annotated list of the grasses of the Uganda Protectorate.* Entebbe: Government Printer.

1956   *The indigenous trees of the Uganda protectorate.* Entebbe: Government Printer.

Ehret, C.

1967   Cattle keeping and milking in eastern and southern African history: The linguistic evidence, *Journal of African History* **8**:1–17.

Eicher, C. K.

1968   Economic research for African agricultural development. In *Conference on agricultural research priorities for economic development in Africa.* Pp. 189–203. Washington, D.C.: National Academy of Sciences.

Eicher, C. K. and L. Witt (editors).

1964   *Agriculture in economic development.* New York: McGraw-Hill.

Ellis, P. J. C.

1957   Ubungu in Chunya District, *Tanganyika Notes and Records* **47/48**: 201–202.

Epstein, H.

1954   The fat-tailed sheep of East Africa, *East African Agricultural Journal* **20**: 109–117.

1955   The Zebu cattle of East Africa, *East African Agricultural Journal* **21**:83–95.

Evans, A.

1960   Studies of intercropping. I. Maize or sorghum with groundnuts, *East African Agricultural Journal* **26**:1–10.

Fagan, B. and J. E. Yellen.

1969   Ivuna: Ancient salt-working in southern Tanzania, *Azania* **3**:1–43.

Firey, W.

1960   *Man, mind, and land: A theory of resource use.* Glencoe, Illinois: Free Press.

Fotheringham, L. M.

1891   *Adventures in Nyasaland.* London: Sampson Low, Marston, Searle and Rivington. (Cited by M. Wilson 1963:2.)

Frake, C. O.

1964   Notes on queries in anthropology, *American Anthropologist* **66** (3, Part 2): 132–145.

Freeman-Grenville, G. S. P.

1963   The German sphere, 1884–98, In *History of East Africa, I,* edited by R. Oliver and G. Matthew. Pp. 433–453. London and New York: Oxford University Press.

Freilich, M.

1963   The natural experiment, ecology, and culture, *Southwestern Journal of Anthropology* **19**:21–39.

Freitag, R. S.

1963   *Agricultural development schemes in sub-Saharan Africa: A bibliography.* Washington, D.C.: Library of Congress.

Fromm, P.
    1912   Ufipa, Land und Leute, *Mitteilungen aus den Deustchen Schutzgebieten*
    25:79–101.
Fuggles-Couchman, N. R.
    1939   Some production cost figures for native crops in the Eastern Province
    of Tanganyika, *East African Agricultural Journal* **4**:396.
    1964   *Agricultural change in Tanganyika 1945–1960*. Stanford, California:
    Stanford University Food Research Institute.
Fülleborn, F.
    1906   *Das Deutsche Njassa- und Ruvuma-Gebiet, Land und Leute*. Berlin:
    Reimer.
Gaitskell, A.
    1968   Importance of agriculture in economic development. In *Economic de-
    velopment of tropical agriculture*, edited by W. W. McPherson. Pp. 46–58. Gains-
    ville: Univerity of Florida Press.
Gann, L.
    1954   The end of the slave trade in British Central Africa: 1889–1912, *Rhodes-
    Livingston Journal* **16**:26–51.

Gardiner, R. K. A.
    1968   Foreword. In *Conference on agricultural research priorities for economic de-
    velopment in Africa*. Pp. vi–vii. Washington, D.C.: National Academy of
    Sciences.

Geertz, C.
    1963   *Agricultural involution: The processes of agricultural change in Indonesia.*
    Berkeley, California: University of California Press.

Gemuseus, O.
    1938   History of Rungwe District, *Tanzania Southern Highlands Provincial Book*,
    entry dated 13 August 1938.

Ghai, D. P.
    1965   *Portrait of a minority: Asians in East Africa*. Nairobi: Oxford University
    Press.
Gibb, Sir A.
    1952   *A development survey for the central African rail link*. London: Sir Alex-
    ander Gibb and Partners.
Gilbert, S. M.
    1945   *The mulching of Arabica coffee*. Nairobi: Government Printer for Tan-
    ganyika Territory Department of Agriculture.
Gillman, C.
    1927   South-west Tanganyika Territory, *Geographical Journal* **69**:97–125.
    1929   *Report on the preliminary surveys for a railway line to open up the south-
    west of Tanganyika Territories*. London: Crown Agents for the Colonies.
    1936   A population map of Tanganyika Territory, *Geographical Review* **26:**
    353–375.
    1938   Problems of land utilization in Tanganyika Territory, *South African
    Geographical Journal* **20**:12–20.
    1949   A vegetation-types map of Tanganyika Territory, *Geographical Review*
    39:7–37.
Gladwin, T.
    1970   *East is a big bird*. Cambridge, Massachusetts: Harvard University Press.
Gleave, M. B.
    1966   Hill settlements and their abandonment in tropical Africa, *Transactions*,
    Institute of British Geographers **40**:39–49.

Glover, J.
1948   Water demands by maize and sorghum, *East African Agricultural Journal* **13**:171.
Gould, P. R.
1969   *Spatial diffusion.* Washington: Association of American Geographers, Commission on College Geography, Resource Paper 4.
Gourou, P.
1956   The quality of land use of tropical cultivators. In *Man's role in changing the face of the earth,* edited by W. L. Thomas. Pp. 336–349. Chicago: University of Chicago Press.
Gras, N. S. B.
1925   *A history of agriculture.* New York: F. S. Crofts.
Great Britain Admiralty.
1916   *A handbook of German East Africa.* Great Britain, Admiralty, Naval Intelligence Division, Geographical Section.
Great Britain Colonial Office
1939   *Report on the administration of Tanganyika Territory for the year 1938.* London: H.M.S.O.
Greenberg, J.
1963   *The languages of Africa.* Bloomington, Indiana: Research Center in Anthropology, Folklore and Linguistics.
Greenway, P. J.
1944   Origins of some East African food plants, *East African Agricultural Journal* **10**:34–39, 115–119, 177–180, 251–256; **11**:56–63.
Greenway, P. J., M. M. Wallace and E. V. R. Khomo.
1953   *The papaw.* Dar es Salaam: Government Printer.
Greig, R. C. H.
1937   Iron smelting in Ufipa, *Tanganyika Notes and Records* **4**:77–81.
Griffiths, I. L.
1968   Zambian links with East Africa, *East African Geographical Review* **6**:87–89.
Grigg, D.
1970   *The harsh lands.* London: Macmillan.
Gulliver, P. H.
1955   *A report on the migration of African workers to the south from the Southern Highlands Province, with special reference to the Nyakyusa of Rungwe District.* Tanganyika Provincial Administration Sociological Research. (Cited by Brock 1966.)
Gunn, D. L.
1956   A history of Lake Rukwa and the Red Locust, *Tanganyika Notes and Records* **42**:1–18.
1957   The story of the International Red Locust Control Service, *Rhodesia Agricultural Journal* **54**:8–24.
Guthrie, M.
1957   *The classification of the Bantu languages.* London: International African Institute.
1969   *Comparative Bantu, An introduction to the comparative linguistics and prehistory of the Bantu language.* Hants,. U.K.: Gregg Press Ltd.
Haarer, A. E.
1928   Letter to Director of Agriculture, Dar es Salaam, dated 10 December 1928. Tanzania National Archives.
1956   *Modern coffee production.* London: Leonard Hill.
Hall, R.
1945   Local migration in Tanganyika, *African Studies* **4**:55–69.

Harkin, D. A.
1960 *The Rungwe volcanics at the northern end of Lake Nyasa.* Dar es Salaam: Government Printer.

Harlow, V. and E. M. Chilver (editors)
1965 *History of East Africa, II.* London and New York: Oxford University Press.

Harman, H. H.
1960 *Modern factor analysis.* Chicago: University of Chicago Press.

Harris, W. V.
1941 Native methods of food storage in Tanganyika, *East African Agricultural Journal* **6**:135–138.

Hartley, B. J.
1936 Groundnuts and their cultivation, *East African Agricultural Journal* **1**: 501–511.

Harwood, A.
1970 *Witchcraft, sorcery, and social categories among the Safwa.* London: International African Institute.

Hatchell, G. W.
1958 The British occupation of the southwestern area of Tanganyika Territory, 1914–1918, *Tanganyika Notes and Records* **51**:131–155.

Henderson, W. O.
1965 German East Africa 1884–1918. In *History of East Africa, II,* edited by V. Harlow and E. M. Chilver. Pp. 123–162. London and New York: Oxford University Press.

Hill, A. G.
1947 Oil plants in East Africa, *East African Agricultural Journal* **12**:140–152.

Hodder, B. W.
1968 *Economic development in the tropics.* London: Methuen.

Hubbell, D. S.
1965 *Tropical agriculture, An abridged guide.* Kansas City, Missouri: H. W. Sams.

Huffnagel, H. P.
1961 *Agriculture in Ethiopia.* Rome: Food and Agriculture Organization of the United Nations.

Huntingford, G. W. B.
1963 The peopling of the interior of East Africa by its modern inhabitants. In *History of East Africa, I,* edited by R. Oliver and G. Matthew. Pp. 58–93. London and New York: Oxford University Press.

Huxley, E.
1953 *White man's country, Lord Delamere and the making of Kenya.* London: Chatto and Windus.

Ingham, K.
1965 Tanganyika: the mandate and Cameron, 1919–1931. In *History of East Africa, II,* edited by V. Harlow and E. M. Chilver. Pp. 543–593. London and New York: Oxford University Press.

Inkles, A.
1970 Becoming modern, *et al.* **2**:58–73.

International Bank For Reconstruction And Development.
1961 *The economic development of Tanganyika.* Baltimore: Johns Hopkins Press.

Iyengar, K. G. *et al.*

1945–1946   Ragi *(Eleusine coracana* Gaertn.), *Mysore Agricultural Journal* **24:** 33–49.

Jacks, G. V., W. D. Brind, and R. Smith

1955   *Mulching.* Farnham Royal, U.K.: Commonwealth Agricultural Bureaux.

Jackson, C. H. N.

1940   Field notes on the species of *Brachystegia* and *Isoberlinia* of Tanganyika Territory, *Journal of South African Botany* 1940: 33–40.

Jacobsen, A.

1951   *Paa Afrikas Vilkaar.* Kφbenhavn: Glydendalske Boghandel.

1954   *Fjorten Afrikanske Aar.* Kφbenhavn: Glydendalske Boghandel.

James, R. W.

1971   *Land tenure and policy in Tangania.* Nairobi and Dar es Salaam: East African Literature Bureau.

Jeffreys, M. D. W.

1953   Pre-Columbian maize in Africa, *Nature* **172:**965–966.

1954   The history of maize in Africa, *South African Journal of Science* **50:** 197–200.

1967   Pre-Columbian maize in southern Africa, *Nature* **215:**695–697.

Johnson, H. B.

1967   The location of Christian missions in Africa, *Geographical Review* **57:** 168–202.

Johnston, B. F. and J. W. Mellor.

1960   The nature of agriculture's contribution to economic development, *Food Research Institute Studies* **1:**335–356.

Johnston, B. F. and H. M. Southworth

1967   Agricultural development: problems and issues. In *Agricultural development and economic growth,* edited by H. M. Southworth and B. F. Johnston. Pp. 1–20. Ithaca, New York: Cornell University Press.

Johnston, H. H.

1890   British Central Africa, *Proceedings, Royal Geographical Society* **12:**713–743.

Jones, W. O.

1957   Manioc: an example of innovation in African economies, *Economic Development and Cultural Change* **5:**97–117.

1959   *Manioc in Africa.* Stanford, California: Stanford University Press.

Joshi, A. B.

1961   *Sesamum.* Hyderabad: Indian Central Oilseeds Committee.

Joshi, N. R., E. A. McLaughlin, and R. W. Phillips

1957   *Types and breeds of African cattle.* Rome: Food and Agriculture Organization of the United Nations.

Kamarck, A. M.

1965   Economics and economic development. In *The African World,* edited by R. A. Lystad. Pp. 221–241. New York: Praeger.

Kay, G.

1965   *Changing patterns of settlement and land use in the Eastern Province of Northern Rhodesia.* Hull, U. K. Hull University Publications, Occasional Papers in Geography No. 2.

Kerr-Cross, D.

1890   Geographical notes on the country between Lakes Nyassa, Rukwa, and Tanganyika, *Scottish Geographical Magazine* **6:**281–293.

King, L. J.
1969   *Statistical analysis in geography*. Englewood Cliffs, New Jersey: Prentice-Hall.
Knight, C. G.
1969   Field work and local government: an example from Tanzania, *African Studies Bulletin* **12**:265–273.
1971a   Ethnogeography and change, *Journal of Geography* **70**:47–51.
1971b   The ecology of Afrian sleeping sickness, *Annals of the Association of American Geographers* **61**:23–44.
1972   The geographer in changing Africa, *Earth and Mineral Sciences* **41**:38–39.
Kootz-Kretschmer, E.
1926–1929   *Die Safwa, ein ostafrikanischer Volksstamm in seinem Leben und Denken*. Berlin: Reimer.
Lee, D. H. K.
1957   *Climate and economic development in the Tropics*. New York: Harper.
Lee, E. C.
1965   *Local taxation in Tanzania*. Berkeley: University of California Institute of International Studies, Research Series No. 6
Lerner, D.
1958   *The passing of traditional society, modernizing the Middle East*. New York: Free Press.
Letcher, O.
1918   Notes on the south-western area of "German" East Africa, *Geographical Journal* **51**:164–172.
Leverett, C. W.
1957   An outline of the history of railways in Tanganyika 1890–1956, *Tanganyika Notes and Records* **47/48**:108–116.
Lind, E. M. and A. C. Tallantire.
1962   *Some common flowering plants of Uganda*. London and New York: Oxford University Press.
Livingstone, I.
1965   The marketing of crops in Uganda and Tanganyika, *African primary products and international trade*, edited by I. G Stewart and H. W. Ord. Pp. 125–147. Edinburgh: Edinburgh University Press.
Lunan, M.
1950   Mound cultivation in Ufipa, Tanganyika, *East African Agricultural Journal* **16**:88–89.
McCulloch, J. S. G.
1965   Tables for the rapid computation of the Penman estimate of evaporation, *East African Agriculture and Forestry Journal* **30**:286–295.
McMaster, D. N.
1962a   *A subsistence crop geography of Uganda*. Bude, U.K. Geographical Publications Limited.
1962b   Speculations on the coming of the banana to Uganda, *Journal of Tropical Geography* **16**:57–69.
McPherson, W. W. (editor)
1968   *Economic development of tropical agriculture*. Gainsville: University of Florida Press.
Majisu, B. N. and N. Doggett.
1968   Some problems of sorghum and millet production in Africa. In *Conference on agricultural research priorities for economic development in Africa*. Pp.112–122. Washington, D.C.: National Academy of Sciences.

Makings, S. M
1967   *Agricultural problems of developing countries in Africa.* London and New York: Oxford University Press.

Makosya, D. A.
1965   Communication techniques in agricultural extension. In *Agricultural development in Tanzania,* edited by H. E. Smith. Pp. 89–93. Dar es Salaam: University College Institute of Public Administration, Study 2.

Malcolm, D. W.
1953   *Sukumaland, an African people and their country.* London and New York: Oxford University Press.

Marcus, E.
1960   Agriculture and the development of tropical Africa, *Land Economics* **36:**172–180.

Masefield, G. B.
1949   *A handbook of tropical agriculture.* London and New York: Oxford University Press.

Mason, I. L. and J. P. Maule.
1960   *The indigenous livestock of eastern and southern Africa.* Farnham Royal, U.K.: Commonwealth Agricultural Bureaux.

Mauny, R.
1953   Notes historique autour des principales plantes cultivées d'Afrique occidentale, *Bulletin d'Institut français d'Afrique Noire* **15:**684–730.

Mbozi Agricultural Office
1968   *Annual Report for 1967–Mbozi District; Agricultural Division.* Typescript.

Mead, W. R.
1953   *Farming in Finland.* London: University of London.

Meister, A.
1968   *East Africa.* New York: Walker.

Mellor, J. W.
1967   Toward a theory of agricultural development. In *Agricultural development and economic growth,* edited by H. M. Southworth and B. F. Johnston, Pp. 21–65. Ithaca, New York: Cornell University Press.

Metze, F.
1967   Personal Communication; District Forest Officer, Ministry of Agriculture, Forests and Wildlife, Mbeya, Tanzania.

Michelmore, A. P. G.
1939   Observations on tropical African grasslands, *Journal of Ecology* **27:**282–312.

Miracle, M. P.
1958   Maize in tropical African agriculture, *Tropical Agriculture* **35:**1–15.
1965   The introduction and spread of maize in Africa, *Journal of African History* **6:**39–55.
1966   *Maize in tropical Africa.* Madison: University of Wisconsin Press.
1967   *Agriculture in the Congo Basin.* Madison: University of Wisconsin Press.

Moffat, U. J.
1932   Native agriculture in the Abercorn District, Northern Rhodesia Department of Agriculture, *Annual Bulletin* **2:**55–62.

Moffett, J. P. (editor)
1958   *Handbook of Tanganyika.* Dar es Salaam: Government Printer.

Montelius, S.
1953   The burning of forest land for the cultivation of crops, *Geografiska Annaler* **35:**41–54.

Morgan, W. B.
1957 "Some comments on shifting cultivation in Africa," *Research Notes, Ibadan University College Department of Geography* **9**:1–10.
1969 Peasant agriculture in tropical Africa. In *Environment and Land Use in Africa*, M. F. Thomas and G. W. Whittington, editors. Pp. 241–272. London: Methuen.
Morgan, W. T. W.
1963 The "White Highlands" of Kenya, *Geographical Journal* **129**:140–155.
Murdock, G. P.
1959 *Africa–Its peoples and their culture history*. New York: McGraw-Hill.
1960 Staple subsistence crops of Africa, *Geographical Review* **50**:523–540.
Mwaliyego, Chief
1967 Personal communication (Mbeya Area, Tanzania).
Mwalupembe, H.
1967 Personal Communication at Rungwe Mission.
Mwamlima, Chief.
1967 Personal Communication (Mbozi Area, Tanzania).
Mwasenga, Chief.
1967 Personal Communication (Mbozi Area, Tanzania).
Napper, D. M.
1965 *Grasses of Tanganyika*. Dar es Salaam: Government Printer.
Nasser, S. F.
1965 Statutary and customary land tenure. In *Agricultural development in Tanzania*, edited by H. E. Smith. Pp. 56–59. Dar es Salaam: University College Institute of Public Administration, Study 2.
Neville-Rolfe, E.
1969 *Economic aspects of agricultural development in Africa*. Oxford: Agricultural Economics Research Institute of the University of Oxford.
Nicholls, L.
1961 *Tropical nutrition and dietetics*. London: Bailliere, Tindall and Cox. (Fourth edition, revised by M. M. Sinclair and D. B. Jelliffe.)
Nowack, E.
1937 Der deutsche Pflanzungsbezirk Tukuyu im südlichen Deutsch-Ostafrika, *Koloniale Rundschau* **28**:393–424.
Nye, P. H and D. J. Greenland.
1960 *The soil under shifting cultivation*. Farnham Royal, United Kingdom: Commonwealth Agricultural Bureau.
Nyerere, J. K.
1962 *Ujamaa–The basis of African socialism*, quotation cited by Ruthenberg 1964; reproduced in Nyerere 1968.
1966 *Principles and development*. Dar es Salaam: Government Printer.
1968 *Ujamaa–Essays on African socialism*. London and New York: Oxford University Press.
Obol-Ochola, J.
1969 Ideology and tradition in African land tenure, *East Africa Journal* **6**:35–41.
O'Connor, A. M.
1965 New railway construction and the pattern of economic development in East Africa, *Transactions, Institute of British Geographers* **36**:21–30.
Ojo, G. J. A.
1968 Some cultural factors in the critical density of population in tropical Africa. In *The population of tropical Africa*, edited by J. C. Caldwell and C. Okonjo. Pp. 312–319. New York: Columbia University Press.

Oldaker, A. A.
1957 Tribal customary land tenure in Tanganyika, *Tanganyika Notes and Records* **47/48**:117–144.
Oliver, R.
1963b Discernible developments in the interior *c.* 1500–1840. In *History of East Africa, I,* edited by R. Oliver and G. Matthew. London and New York: Oxford University Press.
1966 *The missionary factor in East Africa.* London: Longmans Green.
Perham, M.
1931 The system of native administration in Tanganyika, *Africa,* **4**:302–312.
Peters, D. U.
1950 *Land usage in Serenje District.* Rhodes-Livingston Papers 19.
Phillipson, D. W.
1968 The early Iron Age in Zambia: regional variants and some tentative conclusions, *Journal of African History* **9**:191–211.
Pielou, E. C.
1952 Notes on the vegetation of the Rukwa rift valley, Tanganyika, *Journal of Ecology* **40**:383–392.
Pike, A. H.
1938 Soil conservation amongst the Matengo tribe, *Tanganyika Notes and Records* **6**:79–81.
Popplewell, G. D.
1936 Notes on the Fipa, *Tanganyika Notes and Records* **3**:99–105.
Porter, P. W.
1965 Environmental potentials and economic opportunities—a background for cultural adaptation, *American Anthropologist* **67**:409–420.
1968 Personal communication.
Porteres, R.
1951 *Eleusine coracana* Gaertner, cèrèale des humanitès pauvres des pays tropicaux, *Bulletin d'Institut français* d'Afrique Noire **13**:1–78.
1959 Les appelations des cèrèales en Afrique. VII. Les Sorghos (suite), *Journal d'Agriculture Tropicale et de Botanique Appliquée* **6**:68–84
1962 Berceaux agricoles primaires sur le continent Africain, *Journal of African History* **3**:195–210.
Purseglove, J. W.
1968 *Tropical crops: Dicotyledons.* New York: Wiley.
Rainey, R. C., Z. Waloff, and G. F. Burnett
1957 *The behavior of the Red Locust.* London: Anti-Locust Research Centre.
Rattray, J. M.
1954 Some plant indicators in Southern Rhodesia, *Rhodesian Agricultural Journal* **51**:176–186.
Raum, O. F.
1965 German East Africa: changes in African life under German administration, 1892–1914. In *History of East Africa, II,* edited by V. Harlow and E. M. Chilver. Pp. 163–207. London and New York: Oxford University Press.
Raymond, W. D.
1940–1941 The nutritive value of some Tanganyika foods, *East African Agricultural Journal* **6**:105–108, 154–158.
Rhodesia and Nyasaland
1957 Federal Meteorological Department. *Totals of monthly and annual rainfall for selected stations in Northern Rhodesia.* Salisbury: Federal Government Printing and Stationery Department.

Richards, A. I.
   1939   *Land, labour, and diet in Northern Rhodesia.* London and New York: Oxford University Press.
Robertson, J. K.
   1941   Mixed or multiple cropping in native agricultural practice, *East African Agricultural Journal* **6:**228–232.
Robertson, J. K. and C. W. Rombulow-Pearse
   1955   A system of cultivation in the Mbeya District of Tanganyika, *East African Agricultural Journal* **20:**166–167.
Robinson, J. B. D.
   1961   Simple multiple stem pruning of coffee, Coffee Research Station (Lyamungu) *Extension Pamphlet* 5/61.
   1962a   A guide to the conditions of soil and climate suitable for growing Arabica coffee in Tanganyika, Coffee Research Station (Lyamungu) *Extension Pamphlet* 2/62.
   1962b   Arabica coffee production in Tanganyika, *Tanganyika Coffee News* March 1962: 1–12.
   1964   *A handbook on Arabica coffee in Tanganyika.* Moshi, Tanzania: Tanganyika Coffee Board.
Robinson, K. R.
   1966   A preliminary report on the recent archaeology of Ngonde, Northern Malawi, *Journal of African History* **7:**169–188.
Rogers, E.
   1962   *Diffusion of innovations.* New York: Free Press.
Rounce, N. V.
   1949   *The agriculture of the cultivation steppe of the Lake, Western and Central Provinces.* Capetown: Longmans Green.
Rungwe Mission Archives.
   (Rungwe Mission, Rungwe Area, Tanzania).
Rushby, G. G.
   1965   *No more the tusker.* London: W. H. Allen.
Russell, E. W. (editor)
   1962   *The natural resources of East Africa.* Nairobi: East African Literature Bureau.
Ruthenberg, H.
   1964   *Agricultural development in Tanganyika.* Berlin: Springer-Verlag.
   1968   *Smallholder farming and smallholder development in Tanzania.* München: Weltforum Verlag, Afrika Studien 24.
Sauer, C. O.
   1941   The settlement of the humid East. In *Climate and man.* Pp. 157–166. Washington, D.C.: U.S. Government Printing Office.
   1959   Age and area of American agriculture, *Acta 33rd International Congress of Americanists* **1:**181.
Scott, R. M.
   1962   The soils of East Africa. In *The natural resources of East Africa,* edited by E. W. Russell. Pp. 67–76. Nairobi: East African Literature Bureau.
Seddon, D.
   1968   The origins and development of agriculture in East and Southern Africa, *Current Anthropology* **9:**489–494
Simmonds, N. W.
   1959   *Bananas.* London: Longmans Green.

Simoons, F.
1961   *Eat not this flesh.* Madison: University of Wisconsin Press.
Slater, M.
1966   Oedipus and Job in East Africa, *Transactions, New York Academy of Sciences,* Ser. **2, 28:**644–653.
Smith, A.
1963   The southern section of the interior. In *History of East Africa, I,* edited by R. Oliver and G. Matthew. Pp. 253–296. London and New York: Oxford University Press.
Smith, H. E. (editor)
1965   *Agricultural development in Tanzania.* Dar es Salaam: University College Institute of Public Administration, Study 2.
Soja, E. W.
1968   *The geography of modernization in Kenya.* Syracuse: Syracuse University Press.
Southworth, H. M. and B. F. Johnston (editors)
1967   *Agricultural development and economic growth.* Ithaca, New York: Cornell University Press.
Spurr, A. M. M.
1955   *Soils of Mbozi.* Dar es Salaam: Government Printer.
Steel, R. W.
1965   Problems of food and population in tropical Africa, *Kroniek Afrika* **1:** 48–65.
Stevenson, D.
1931   Some important native timbers, Northern Rhodesia Department of Agriculture *Annual Bulletin* **1:**43–52.
Stewart, J.
1881   Lake Nyassa, and the water route to the Lake Region of Africa, *Proceedings, Royal Geographical Society,* New Ser. **3:**257–277.
Sturtevant, W. C.
1964   Studies in ethnoscience, *American Anthropologist* **66,**3 Part 2:99–131.
Swaine, G.
1961   *Insect pests of maize.* Dar es Salaam: Government Printer.
Swynnerton, R. J. M.
1966   Natural resources that govern the future, *African Affairs* **65:**40–54.
Tanganyika ARPC.
Tanganyika Territory. *Annual Reports of the Provincial Commissioners on Native Administration for the Year –.* Dar es Salaam: Government Printer. (For years 1929–1959, dated 1930–1960, respectively.)
Tanganyika Department of Agriculture.
1929   *Annual report 1928.* Dar es Salaam: Government Printer.
1933   *Annual report 1932.* Dar es Salaam: Government Printer.
1935   *Annual report 1934.* Dar es Salaam: Government Printer.
1945   *Agriculture in Tanganyika.* Dar es Salaam: Government Printer.
Tanganyika. Engineering Geologist.
1955   Report on the hydrologic conditions of the Mbozi cultivated area— Mbeya District, Tanzania Archives File Acc 135 25/1 (Dar es Salaam).
Tanganyika Legislative Council.
1958   Review of land tenure policy, *Government Paper* No. 6.
Tanganyika. MbDR
1960   *Mbozi District report.* (Vwawa, Tanzania).

Tanganyika. MDR
  Mbeya District Reports. Typescript. National Archives, Dar es Salaam.
Tanganyika Nonnative Census
  1958 *Report on the census of the non-African population taken on the night of 20th/21st February, 1957.* Dar es Salaam: Government Printer.
Tanner, R. E. S.
  1951. Some Southern Province trees with their African names and uses, *Tanganyika Notes and Records* **31**:61.
Tanzania. Archives Files
  Acc 135 25/1: Mbeya District (Dar es Salaam).
  Acc 157 L2/11: Native Law and Custom, Land Tenure (Dar es Salaam).
  33// L2/45: Initial Applications for Land: Unyiha (Dar es Salaam).
  77//2/33: Iringa Province, Agriculture and Soil Erosion (Dar es Salaam).
  11039: Secretariat (Dar es Salaam).
  40511/11: Projects: Ranching (Dar es Salaam).
Tanzania.
  IPB. *Iringa Provincial Book.* National Archives: Dar es Salaam.
Tanzania.
  MDB. *Mbeya District Book.* National Archives: Dar es Salaam.
Tanzania.
  MbDB. *Mbozi District Book.* National Archives: Dar es Salaam.
Tanzania.
  SHPB. *Southern Highlands Provincial Book.* National Archives: Dar es Salaam.
Tew, M.
  1950 *Peoples of the Lake Nyasa region.* London: Oxford University Press for the International African Institute.
Thomas, I. D.
  1968a Geographical aspects of the Tanzania population census 1967, *East African Geographical Review* **6**:1–12.
  1968b Population density in Tanzania, 1967, University College-Dar es Salaam, Bureau of Resource Assessment and Land Use Planning *Research Notes* No. 5b.
  1972 Infant mortality in Tanzania, *East African Geographical Review* **10**:5–26.
Thomson, J.
  1881 *To the central African lakes and back.* London: Sampson, Low, Marston, Searle and Rivington.
Thornthwaite, C. W.
  1948 An approach toward a rational classification of climate, *Geographical Review* **38**:85–94.
Thwaites, D. H.
  1944 Wanyakyusa agriculture, *East African Agricultural Journal* **9**:236–239.
Trapnell, C. G.
  1943 *The soils, vegetation, and agriculture of northeastern Rhodesia: Report of the Ecological Survey.* Lusaka: Government Printer.
  1959 Ecological results of woodland burning experiments in northern Rhodesia, *Journal of Ecology* **47**:129–168.
Trapnell, C. G. and J. N. Clothier.
  1937 *The soils, vegetation and agriculture of northwestern Rhodesia: Report of the Ecological Survey.* Lusaka: Government Printer.
Trapnell. C. G. and I. Langdale-Brown.
  1962 The natural vegetation of East Africa. In *The natural resources of East Africa,* edited by E. W. Russell. Pp. 92–102. Nairobi: East African Literature Bureau.

Tyler, S. A.

1969   *Cognitive anthropology.* New York: Holt, Rinehart and Winston.

Vernon, M.

1968   Personal communication.

Vesey-Fitzgerald, D. F.

1955   The vegetation of the outbreak areas of the Red Locust (*Nomadacris septemfasciata* Serv.) in Tanganyika and Northern Rhodesia, *Anti-Locust Bulletin* **20**. London: Anti-Locust Research Centre.

Waibel, L.

1950   European colonization in southern Brazil, *Geographical Review* **40**: 529–547.

Wakefield, A. J.

1933   *Arabica coffee, periods of growth and seasonal measures.* Dar es Salaam: Government Printer.

Watson, W.

1958   *Tribal cohesion in a money economy.* Manchester: Manchester University Press for the Rhodes-Livingstone Institute.

Watters, R. F.

1960   The nature of shifting cultivation: a review of recent research, *Pacific Viewpoint* **21**:59–99

Webster, C. C. and P. N. Wilson

1966   *Agriculture in the tropics.* London: Longmans Green.

Wenban-Smith, W.

1963   The Germans on Lake Nyasa, *Tanganyika Notes and Records* **61**:217–218.

Wharton, C. R.

1968   Risk, uncertainty, and the subsistence farmer. Paper read at the Joint Meeting of the American Economic Association and the Association of Comparative Economics, Chicago. Partially reprinted, *Economic development and social change,* edited by G. Dalton. Pp. 566–574. New York: The Natural History Press.

White, F.

1962   *Forest flora of Northern Rhodesia.* London and New York: Oxford University Press.

Whyte, R. O.

1962   The myth of tropical grasslands, *Tropical Agriculture* **39**:1–11.

Willis, J.

1967   Personal communication.

Willis, J. C.

1909   *Agriculture in the tropics.* Cambridge, England: Cambridge University Press.

Willis, R. G.

1964   Traditional history and social structure in Ufipa, *Africa* **34**:340–351.

1966   *The Fipa and related peoples of south-west Tanzania and north-east Zambia.* London: International African Institute.

1968   The Fipa. In *Tanzania before 1900,* edited by A. Roberts. Pp. 82–95. Nairobi: East African Publishing House.

Wills, A. J.

1964   *An introduction to the history of Central Africa.* London and New York: Oxford University Press.

Wilson, M.

1958   *Peoples of the Nyasa-Tanganyika Corridor.* Capetown: School of African Studies, University of Capetown.

Wilson, M.
1963   *Good company, A study of Nyakyusa age-villages.* Boston, Massachusetts: Beacon Press.
Wilson, S. G.
1941   Agricultural practices among the Angoni–Tumbuka tribes of Mzimba (Nyasaland), *East African Agricultural Journal* **7**:89–93.
Winter, E. H.
1955   *Bwamba economy.* Kampala: East African Institute Social Research.
von Wissman, H.
1891   *My second journey through Equatorial Africa.* London: Chatto and Windus.
Wolfe, H.
1928   European settlement areas in the Iringa Province, *East Africa* 16 February 1928:726–728.
Wood, A.
1959   *The groundnut affair.* London: Faber and Faber.
Wright, M.
1968a   Chief Merere and the Germans, *Tanzania Notes and Records* **69**:41–49.
1968b   Local roots of policy in German East Africa, *Journal of African History* **9**:621–630.
1971   *German missions in Tanganyika 1891–1941.* Oxford: Clarendon Press.
Wrigley, G.
1961   *Tropical agriculture.* London: B. T. Batsford.
Yaeger, R.
1968   *Personal communication.*
Yeoman, G. H
1956   *The occurrence of East Coast Fever in Tanganyika.* Dar es Salaam: Government Printer.
Yudelman, M.
1964   *Africans on the land.* Cambridge, Massachusetts: Harvard University Press.

Index and Glossary

A 4
B 5
C 6
D 7
E 8
F 9
G 0
H 1
I 2
J 3